INTERCONNECT NOISE OPTIMIZATION IN NANOMETER TECHNOLOGIES

INTERCONNECT NOISE OPTIMIZATION IN NANOMETER TECHNOLOGIES

Mohamed A. Elgamel and Magdy A. Bayoumi
The Center for Advanced Computer Studies
University of Louisiana at Lafayette
P.O.Box 44330
Lafayette, LA 70504, USA

 Springer

Mohamed A. Elgamel and Magdy A. Bayoumi
The Center for Advanced Computer Studies
University of Louisiana at Lafayette
P.O. Box 44330
Lafayette, LA 70504 U.S.A.

Interconnect Noise Optimization in Nanometer Technologies

e-ISBN 0-387-29366-3

ISBN 978-1-4419-3844-2   eISBN 978-0-387-29366-0

Printed on acid-free paper.

9 8 7 6 5 4 3 2 1

springeronline.com

To my family, M. Elgamel

To my students, M. Bayoumi

Contents

LIST OF FIGURES

LIST OF TABLES

Preface

As the transistor feature size keeps shrinking, many new challenges are arising in designing current and future microelectronic systems. According to the International Technology Roadmap for Semiconductors (ITRS), the feature size of a transistor has been scaled down from 130nm in 2001 to 70nm in 2005, and it is expected to be further reduced to 45nm in 2010. One of the main challenges in these submicron technologies is interconnect among million devices on the same chip. Wires are placed closer to each other, which leads to larger coupling capacitance and inductance. The number of interconnect levels and the amount of interconnect increase considerably. The interconnect delay dominates over gate delay in determining circuit performance.

These new design issues, parameters, and challenges lead to significant increase in design complexity that requires new design tools and methods. There are not many design tools that address all these issues in an integrated and comprehensive approach. Moreover, there are only very few books that address the new design challenges in submicron (nano) technologies.

This monograph addresses some of these important issues. It provides insight and intuition into layout analysis and optimization for interconnect in high speed, high complexity integrated circuits in nanotechnologies. It brings together a wealth of information previously scattered throughout the literature, presenting a range of CAD algorithms and techniques for synthesizing and optimizing interconnect. It helps to fill the existing gap in this area. The monograph can be used as a text book and/or a supplementary book in research and advanced courses in Nanotechnologies, Interconnect Design, CAD Tools, and Design Methodologies. It can also serves as a material for tutorials and short courses in related areas.

In this monograph, the effects of wire width, spacing between wires, wire length, coupling length, load capacitance, rise time of the inputs, place of overlap (near driver or receiver side), frequency, shields, and direction of the signals on system performance and reliability are thoroughly investigated. In addition, parameters like driver strength have been considered as several recent studies considered the simultaneous device and interconnect sizing. Crosstalk noise, as well as the impact of coupling on aggressor delay, is analyzed. The pulse width of the crosstalk noise, which is of similar importance for circuit performance as the peak amplitude, is also analyzed.

The practical aspects of the algorithms and models are explained with sufficient detail. It investigates in-depth the most effective parameters in layout optimization that can affect both capacitive and inductive noise. Noise models needed for layouts with multi-layer multi-crosscoupling segments are investigated. Different post-layout optimization techniques are explained with complexity analysis and benchmarks tests are provided. A list of state of the art CAD tools available at this time is included, highlighting the practical use for each one. Also, a chapter for the 3D integration technologies and the 3D design tools issues is included.

Acknowledgments

The authors acknowledge the support of the U.S. Department of Energy (DoE), EETAPP program, DE97ER12220 and the Governor's Information Technology Initiative. The environment in the Center for Advanced Computer Studies (CACS) has been dynamic, inspiring, and supportive for such project. The authors would also like to warmly thank the Springer staff for their support and assistance. We deeply appreciate the Edith Garland Dupré Library Director for providing the University of Louisiana at Lafayette with IEEE Xplore service. This service was very vital for this work. Dr. Elgamel would like to thank the Arab Academy for Science and Technology, AAST, and its president Dr. Gamal Mokhtar, for their support. He is particularly thankful to Dr. Yasser Hanafy, Dean of College of Computing and Information Technology. Dr. Bayoumi would like to thank his students, former and current, for enriching his life. they are keeping him young in heart and spirits. They are making the academic life exciting, interesting, and never boring.

Chapter 1

INTRODUCTION

The dramatic scaling of integrated circuit technologies over the last two decades has allowed active device counts to reach hundreds of millions. The interconnect among the devices tends to grow super linearly with the transistor counts. Interconnect delay will dominate the total circuit delay with further technology scaling. To provide more routing resources for the increased amount of interconnect, numerous metal layers are accommodated and diagonal routing, in addition to Manhattan routing, are frequently used. Also, 3D technology is becoming of more interest to the semiconductor electronic industry. This increased interconnect create problems with the signal integrity and interconnect delay. We introduce some of these problems and the interconnect models used to represent these problems. We also introduce some solutions and analysis techniques to alleviate these problems.

The goal of this chapter is to introduce the technology trends and the interconnect scaling with technology and provide the readers with an overview of the motivations and the contents of this book.

1.1 TECHNOLOGY TRENDS

As reported by the International Technology Roadmap for Semiconductors (ITRS)[1], the feature size of VLSI devices was scaled down from 130nm in 2001, to 70nm, and is expected to be further reduced to 45nm in 2010. Some of the main characteristics of each technology generation according to ITRS is listed in Table 1-1.

Table 1-1. Technology trend according to ITRS

Year	2001	2002	2003	2004	2005	2006	2007
Technology (nm)	150	130	107	90	80	70	65
Number of metal levels	8	8	8	9	10	10	10
Total interconnect length (m/cm^2)	4086	4843	5788	6879	9068	10022	11169
Height/width (aspect ratio)	1.6	1.6	1.6	1.7	1.7	1.7	1.7
On-chip clock frequency (GHz)	1.6	2.3	3	4	5	5.6	6.7
Dielectric constant	<2.7	<2.7	<2.7	<2.4	<2.4	<2.4	<2.1
Min. wire width	150	130	107	90	80	70	65
Min. wire spacing	210	170	140	120	105	100	85

1.2 MOTIVATION

The current trend of technology scaling is predicted to continue. Feature sizes will continue to shrink to very deep submicrometer (VDSM) dimensions while clock frequencies will increase. Many new challenging issues arise. First, much smaller and faster devices exist, but interconnect delay dominates over gate delay in determining circuit performance. If devices and interconnects are all scaled down by a factor S, the intrinsic gate delay will decrease by a factor of S, and the delay of local interconnects (connecting adjacent gates) remains the same[2], but the delay of global interconnects increases by a factor of S^2.

Second, with technology scaling, wires are placed closer to each other and the aspect ratio increases. This leads to larger coupling capacitance, which causes crosstalk noise and excessive signal delay. Third, high clock frequencies, faster transistor rise/fall time, long signal wires, and the use of wider wires and Cu material interconnects, inductance of interconnect and the noise generated because of this inductance are becoming important design metrics to be considered in digital designs. Fourth, scaling down the feature size allows more transistors to be integrated on the chip. The number of interconnect levels and the amount of interconnect to be dealt with will increase with the advance of technology. This leads to significant increase in design complexity, which requires new design tools that can deal with the new challenges. All these effects have made interconnect design one of the most challenging problems for high-performance IC design.

1.3 BOOK OUTLINE

This book is organized in the following manner. Chapter 2 lays the groundwork for this book. In this chapter, we introduce the definition of noise and some of the noise sources that are encountered in VLSI circuits. We also elaborate on some of the existing noise reduction techniques.

Chapter 3 presents some interconnect models and categorizes them. It also introduces some interconnect noise minimization techniques that are required in the nanometer technologies. In chapter 4, we analyze the effects of all known interconnect and driver parameters on the crosstalk peak noise, crosstalk noise pulse width, and the impact of coupling on aggressor delay. We consider parameters like spacing between wires, wire length, coupling length, load capacitance, rise time of the inputs, place of overlap (near driver or receiver side), frequency, shield insertion, direction of the signals, and wire width for both the aggressors and the victim wires. Also, we consider parameters like driver strength as several recent studies considered the simultaneous device and interconnect sizing. We use the simulation results as guidelines on how to optimize interconnect and which interconnect parameter should be given high priority for optimization. We consider inductance effects on interconnect and perform experiments to report parameters affecting inductance and the noise generated due to this inductance. We also report how effective each of these parameters are. In chapter 5, we formulate the min-area shield insertion problem and provide an efficient solution that satisfies given explicit noise bounds in multiple coupled nets. Chapter 6 introduces several available algorithms that use the spacing technique as a tool to decrease interconnect crosscoupling noise. Some of these algorithms consider spacing alone and others consider spacing and other techniques, like sizing, at the same time. In chapter 7, we propose a new framework that minimizes the crosstalk between adjacent wires. We limit our work to gridless routed 2-layer layouts for simplicity. However, the approach can be applied to a multi-layer routing. In this framework, we sort the nets and the segments inside each net in descending order of crosstalk. Through an iterative approach, we minimize the maximum crosstalk without allowing any other net to have a new crosstalk value greater than the one being optimized. We expand a specific noise model to handle multi segment nets with multiple crosscoupling. We extend the crosstalk noise minimization methods to account for the changes of crosscoupling and wire segments lengths in other layers. Chapter 8 introduces the 3D technology and the already available commercial devices using this technology. We survey some of the available 3D integration technologies and the 3D IC design tools. Finally, in chapter 9 we survey some tools in the EDA industry and outline their features and key benefits.

Chapter 2

NOISE ANALYSIS AND DESIGN IN DEEP SUBMICRON

Traditionally, area-minimization and speed-maximization were the only factors relative to a design's effectiveness that were measured. Low power, high-throughput, and computationally intensive circuits are also critical application domains[1]. In addition to these three design parameters; area, speed, and power, there are two design metrics, which have been of great importance to current designs. These metrics are noise and reliability[3,4]. The five metrics are shown in Fig. 2-1 with an arrow associated to show whether this metric should be increased or decreased.

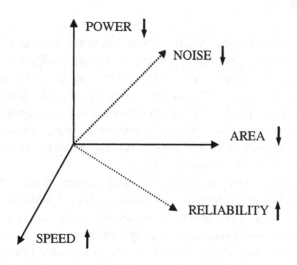

Figure 2-1. Design metrics.

The current trend of technology scaling is predicted to continue. Feature sizes will continue to shrink to very deep submicrometer (VDSM) dimensions while clock frequencies will continue to increase. Shrinking feature size implies not only shorter gate lengths but also decreasing interconnect pitch and device threshold voltages. Now, a single IC can contain an entire system (a system-on-a-chip, SOC), and the interconnections can have many interleaved signal layers and multiple planes of interconnect. Reduction in the top and bottom areas of a minimum-width wire means that total wire capacitance is decreasing. Resistance, however, is increasing faster, despite efforts not to scale metal thickness. Practical efforts to control RC delays through the use of low-resistivity metal (copper), low-dielectric-constant insulators, and wide, thick wiring will require future interconnection analysis to consider inductance and inductive coupling. It is projected that use of a lower resistivity metal (copper) and that replacement of silicon dioxide (k~4) with various insulating materials of a progressively lower dielectric constant (k~2–3) will be adopted in the chip fabrication process. The use of copper will reduce degradation of signal propagation delay time due to voltage drop on power lines. The use of low k dielectrics will decrease the degradation of signal propagation delay time by reducing capacitive coupling. While technology scaling results in lower threshold voltages, the threshold voltage magnitude determines noise immunity in these circuits.

2.1 NOISE

The term noise in digital VLSI systems has come to mean any unwanted deviation in the voltages and/or currents at various nodes within a circuit. When noise acts against a stable logic level on a circuit node, it can transiently destroy logical information carried by the node. If this ultimately causes an incorrect machine state stored in a latch, functional failure will result. Even when noise does not cause functional failure, it has an impact on timing, affecting both delay and slew. Also, noise can cause the dissipation of extra power due to incorrect switching.

Digital circuits create deterministic noise several orders of magnitude greater than noise from stochastic physical sources. Problems due to these noise sources were first observed in mixed signal applications, which plunged highly noise-sensitive analog circuits into a noisy digital environment. Although digital circuits create much more noise than analog circuits, digital systems are prevalent because they are inherently immune to noise. Valid voltage ranges for defining the digital 0 and 1 are shown in Fig. 2-2.

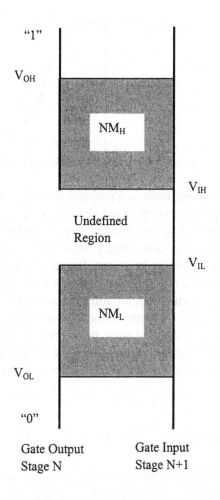

Figure 2-2. A range of analog voltages defines the digital 0 and 1.

The inherent noise immunity of digital circuits is due to the presence of high-gain restoring logic gates such as the inverter shown in Fig. 2-3, which has a very nonlinear voltage transfer characteristic. However, as power supply levels have decreased, this advantage has diminished. Thus, the problem of noise has increased in importance such that on-chip noise is the main research area for continuing the growth in integrated circuit density and performance.

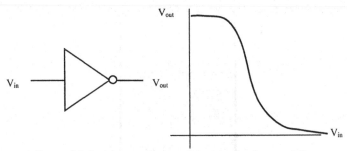

Figure 2-3. Inverter nonlinear voltage transfer characteristics.

With technology scaling, noise is a problem affecting all types of designs from custom microprocessors to standard-cell ASICs. A noise analysis solution must be capable of analyzing tens of millions of transistors, considering both circuit and interconnect noise, and evaluating the distinct noise tolerances of each node in the circuit. Successful design methodologies incorporate a three-level noise strategy. The first line of defense is a set of noise avoidance rules to guide circuit and interconnect design. These rules should prevent most noise problems without introducing too much area or timing constraints. Next, a detailed static noise analysis of the design should find all possible noise failures. Finally, careful circuit simulation should determine whether the design could tolerate some failures flagged by static noise analysis.

2.2 RELIABILITY

One of the most important attributes of any system is its reliability. It is imperative to consider reliability into complex chips even more carefully as the chip functionality increases almost without limit. This important issue is addressed with emphasis on how to consider reliability from early design phases[4], rather than treating reliability assurance as a backend manufacturing process. As CAD tools have played key roles in developing integrated circuits and systems, new reliability analysis tools need to be developed and used in SOC design. For more than a decade, new CAD capabilities have been developed with reliability focus, specifically to address reliability concerns due to hot carrier induced degradation of circuits, electromigration, ESD, crosstalks, leakage currents and high power dissipation.

It was also reported in ITRS 1999 Edition 5 that computer-aided design (CAD) tools would need to incorporate contextual reliability considerations in the design of new products and technologies. It is essential that advances in failure mechanism understanding and modeling, which result from the use of improved test methodologies, be used to provide input data for these new CAD tools. With these data and smart reliability CAD tools, the impact on product reliability of design selections can be evaluated. New CAD tools need to be developed that can calculate degradation in electrical performance of the circuit over time. The inputs used would be the predicted resistance increases in interconnect wires and vias in the circuit based on the following:

- wire length;
- current densities;
- calculated local operating temperature, which includes the effects of Joule heating in the circuit element and elsewhere.

These specific tools will need to become an integral part of the circuit designer's tool set. They will help to predict product reliability before processing begins and to develop solutions that anticipate technology and thereby, accelerate their introduction.

2.3 NOISE SOURCES

Serious on-chip electrical problems are being encountered in DSM. These problems include signal distortion along coupled interconnect lines, voltage variations in the power supply distribution, substrate coupling, charge sharing, charge leakage, process variation, thermal noise, and alpha particles, each of which is a major source of on-chip noise in VLSI circuits.

2.3.1 Interconnect crosscapacitance noise

Interconnect Crosscapacitance noise refers to the charge injected in quiet wires (victims) by neighboring switching wires (aggressors) through the capacitance between them (Crosscapacitance). The resulting noise has the form of a pulse in which the leading edge is determined by the switching slew of the aggressor and the trailing edge is determined by the restoring time constant of the victim. This is perceived to be the most significant source of noise in current processes, Fig. 2-4. If the net is a dynamic node, the restoring time is infinite and the node will never recover. It can lead to setup violations in downstream latches or flip-flops. Also when the aggressors are switching in the same direction of the victim, the transport delay and slew decreases, leading to hold time violations.

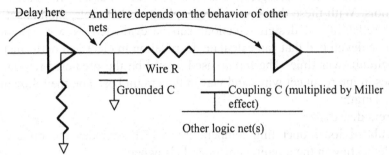

Figure 2-4. Various noise sources for digital circuits.

2.3.2 Charge sharing noise

Charge sharing noise is caused by charge redistribution between a dynamic evaluation node and intermediate nodes in pull up or pull down logic stack, Fig. 2-5. This primarily impacts domino nodes, weakly driven pass gate latches, and dynamic latches. The primary technology variable here is the ratio of junction capacitance to gate and interconnect capacitance. For most circuits, this noise is not dramatically changed with technology scaling.

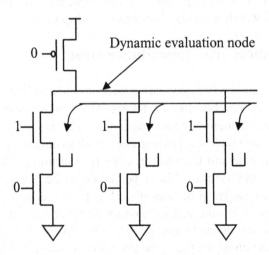

Figure 2-5. Charge sharing noise

2.3.3 Charge leakage noise

Charge leakage noise is mainly composed of subthreshold conduction in nominally off transistors. The leakage current can either charge/discharge a dynamic node or cause the stable state of a weakly held node to be significantly different from rails. This is mainly a concern for wide domino NOR, PLA, and memory arrays. This current rises exponentially with reduction in the threshold voltage and is becoming very significant in DSM. It is helped greatly by feedback devices.

2.3.4 Power supply noise

Power supply noise is the difference between the local voltage references of the driver and the receiver. The increased amount of current on power supply lines causes a raise in IR drop on voltage references. This makes the gate more highly sensitive to noise spikes. This low frequency component (IR drop) is managed well by flip-chip C4 packaging, which provides a very low resistance current path. Besides, the higher speed transients allowed by scaled transistor sizes is associated with higher $L\ di/dt$ due to the large package and on chip inductances. Furthermore, the decoupling capacitance of the circuit is decreased due to the reduced sizes of the gates. Power supply noise is a dominant factor in the design of wide domino circuits and in circuits using contention where the AC logic level is shifted with respect to power supply rails. To counter these problems in high speed designs, several physical design techniques have been proposed: sizing up the P/G lines to accommodate the large current peaks and to minimize the IR and $L\ di/dt$ voltage variations in these lines[6]; increasing the number of P/G pins; and deploying decoupling capacitors in the P/G lines[2,7]; performing clock skew scheduling to minimize the number of simultaneous switching[8]; and using copper in place of aluminum to overcome the increased resistance of scaled interconnect.

2.3.5 Mutual inductance noise

Mutual inductance noise occurs when signal switching causes transient current to flow through the loop formed by the signal wire and current return path[9], thereby creating a changing magnetic field, Fig. 2-6. This induces a voltage on a quiet line, which is in or near this loop. These noise sources can be cumulative if there are several signals switching simultaneously in a bus. Mutual inductance is a long-range phenomenon and, hence, is worse in the presence of wide busses. High-speed switching and synchronous bus structures are making this noise very significant in current technologies.

Inductive noise can combine with capacitive noise to cause even worse noise than shown in Fig. 2.6. Because the analysis of inductive effects is highly dependent on layout and is quite complex, the approach is usually to solve the problem out through rules rather than analyze arbitrary configurations.

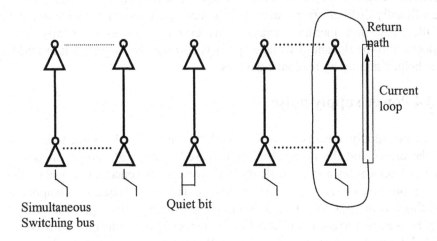

Figure 2-6. Mutual inductance noise from simultaneous switching on a wide bus.

2.3.6 Thermal noise

Thermal effects are an inseparable aspect of electrical power distribution and signal transmission through the interconnects due to self-heating (or Joule heating) caused by the flow of current. Thermal effects impact interconnect design and reliability in the following ways: First, they limit the maximum allowable RMS current density (since the RMS value of the current density is responsible for heat generation) in the interconnects, in order to limit the temperature increase. Second, interconnect lifetime (reliability), which is limited by electromigration (EM) (transport of mass in metals under an applied current density), has an exponential dependence on the inverse metal temperature. Hence, temperature rise of metal interconnects, due to self-heating phenomena, can also limit the maximum allowed average current density, since EM capability is dependent on the average current density[10]. Third, thermally induced open-circuit metal failure under short-time high peak currents (including ESD) is also a reliability concern[11] and can introduce latent EM damage that has important reliability implications[12]. It has been argued that thermal effects will increasingly dominate interconnect design rules that specify maximum current densities for circuit designers[13]. Recently, Hunter[14] has followed up on this issue by solving the EM lifetime equation for Al-Cu, and the 1D heat equation in a self-consistent manner. In this approach, both EM and self-heating can be comprehended simultaneously.

2.3.7 Process variation

The typical characteristics of the process like the gate oxide can vary among wafers or even on a single die. The devices and their properties are defined only within a certain margin and hence will affect the performance of the circuit.

2.4 NOISE REDUCTION TECHNIQUES

Noise reduction techniques can be classified into four categories: signal-encoding schemes, which have been proposed to minimize transition activities on buses; circuit techniques to make circuits more immune to noise; interconnect structures techniques, which are changing the interconnect topology, wire sizing, spacing, and buffer locations; and high-level synthesis techniques. In addition to these categories, using new materials in interconnect including fiber optics or electromagnetic transmission, 3D interconnects which uses multiple levels of active devices, and new packaging methodologies[1] are some other techniques.

The most common techniques for reducing noise in digital circuits include disallowing of

- Pass gates at the ends of long wires
- Long wire runs feeding domino gate inputs
- Single n-FET or p-FET pass gates because of the V_t voltage drop they cause
- High-beta static circuits feeding low beta static circuits or vice-versa.

2.4.1 Signal encoding techniques

Increased coupling effect between interconnects in ultra deep sub-micron technology not only aggravates the power-delay metrics but also deteriorates the signal integrity due to capacitive and inductive crosstalk noises. Conventional approaches to interconnect synthesis aim at optimal interconnect structures in terms of interconnect topology, wire width and spacing, and buffer location and sizes[15].

Signal encoding schemes have been proposed to minimize transition activities on buses while ignoring cross-coupled capacitances. When statistical properties are unknown a priori, the bus-invert method[16] and the on-line adaptive scheme[17] can be applied to encode randomly distributed signals. On the other hand, highly correlated access patterns exhibit a spatio-temporal locality, which can be exploited for energy reduction[18] in Gray code[19,20], the T0 method[21], the working-zone encoding[22], the combined bus-invert/TO[23], and the coupling-driven method[24]. Lower bounds for minimum achievable transition activity have been derived for noiseless buses[25] and for noisy buses[26]. A segmentation method was introduced to reduce power consumption[27]. In specification, transformation approaches were used to reduce the number of memory accesses at the behavioral level[28]. The effectiveness of various encoding schemes was compared at the system level[29].

2.4.1.1 Bus-invert encoding

The bus-invert coding has been introduced to reduce the bus activity: Hamming distance between the consecutive binary numbers. If the Hamming distance of the two consecutive binary numbers is more than half of the word length, the latter binary number is sent in inverted polarity by asserting an additional signal line that indicates bus inversion[16]. It can be used to reduce the weight (the number of ones or zeros) of the binary numbers if the bus-inversion decision is made when the weight is more than half of the bus width. The bus-invert method is as follows:

1. Compute the Hamming distance (the number of bits in which they differ) between the present bus value (also counting the present invert line, Fig. 2-7) and the next data value.
2. If the Hamming distance is larger than n/2, set invert = 1 (and thus make the next bus value equal to the inverted next data value).
3. Otherwise, let invert = 0 (and let the next bus value equal to the next data value).
4. At the receiver side, the contents of the bus must be conditionally inverted according to the invert line, unless the stored data are not encoded as it is (e.g., in a RAM). In any case, the value of invert must be transmitted over the bus (the method increases the number of bus lines from n to n + 1).

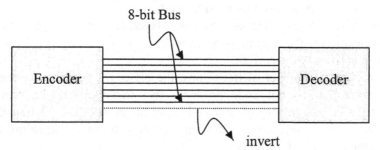

Figure 2-7. Invert signal in bus-invert method.

2.4.1.2 The Gray code encoding

Gray code has only one-bit different in consecutive numbering for addressing. Due to the locality of program execution, Gray code addressing can significantly reduce the number of bit switches. Experimental results show that for typical programs running on a RISC microprocessor, using Gray code addressing reduces the switching activity at the address lines by 30~50% compared to using normal binary code addressing.

2.4.1.3 The TO encoding

The bus transitions are reduced by freezing the address lines when consecutive patterns are found to be sequential. An extra bus line is employed to inform the receiver side whether or not the current pattern is sequential.

2.4.1.4 The working-zone encoding (WZE)

The basis of the WZE technique is as follows:
1. It takes into account the locality of the memory references: applications favor a few working zones of their address space at each instant. In such cases, a reference can be described by an identifier of the working zone and by an offset. This encoding is sent through the bus.
2. The offset can be specified with respect to the base address of the zone or to the previous reference to that zone. Since we want small offsets encoded in a one-hot code, the latter approach is the most convenient. As a simple example, consider an application that works with three vectors (A, B, and C) as shown in Fig. 2-8. Memory references are often interleaved among the three vectors and frequently close to the previous reference to the vector. Thus, if both the sender and the receiver had three

registers (henceforth named holding a pointer to each active working zone, the sender would only need to send:

1. the offset of the current memory reference with respect to the association to the current working zone;
2. an identifier of the current.

To reduce the number of transitions, the offset is encoded in a one-hot code. Since the one-hot code produces two transitions if the previous reference was also in the one-hot code, and an average of n/2 transitions when the previous reference is arbitrary, using a transition-signaling code reduces the number of transitions.

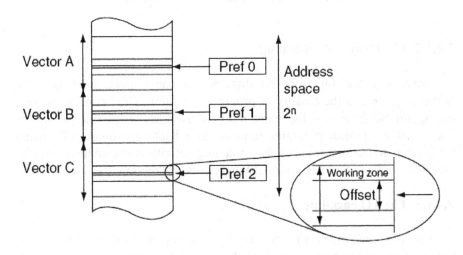

Figure 2-8. Address space for three vectors.

2.4.1.5 Coupling-driven signal encoding

The key idea is that transforming the signal sequences traveling on-chip buses that are closely placed could alleviate coupling effects[24]. Small blocks of encoding and decoding logic are employed at the transmitter and receiver of on-chip buses as shown in Fig. 2-9.

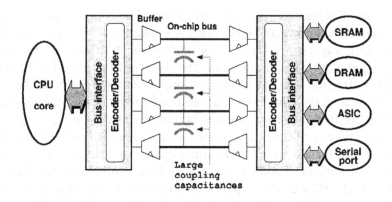

Figure 2-9. Tightly cross-coupled on-chip buses in a system-level chip design.

There are four types of possible transitions when dynamic charge distribution is considered over coupling capacitances as in Fig. 2-10. There are two parallel wires placed with minimum spacing. A type I transition occurs when one of the signals switches while the other stays unchanged such that the coupling capacitance is then charged up to $k_1 C_x V$, where the coefficient $k1$ is introduced as a reference for other types of transition. In a type II transition, one bus switches from low to high while the other switches from high to low. The effective capacitance will be larger than k_1 by a factor of k_2 the value of which is usually two. In a type III transition, both signals switch simultaneously and C_x will not be charged. However, because of possible misalignment of the two transitions, the amount of power consumption varies according to the dynamic characteristics by a factor of k3. In a type IV transition, there is no dynamic charge distribution over coupling capacitance. Thus, k_4 is set to zero.

There are some assumptions. First, synchronous latches are located at the transmitter side, thus all the transitions shall take place at the same time on the bus. The simultaneous transitions exclude type III transitions by setting $k_3 = 0$. It means that the achieved results are on the lower end of power saving. Second, statistics on the information source are not given in advance. Hence this scheme is suitable for data bus encoding, where it is difficult to extract accurate probabilistic information off-line. Enumeration method is employed to represent the coupling effect. If a bus line Bi is located between two other lines, a signal transition on B_i can trigger charge shifts on both coupling capacitances connected to B_{i-1} and B_{i+1}, respectively. In other words, at most two couplings can be initiated by a signal transition. Thus, $2(N-1)$ bits are sufficient to represent the whole set of couplings in an N-bit bus per bus cycle. According to the types of correlated transition between neighboring buses, the coupling encoder generates a codeword as follows: 00 for a type III or IV transition, 01 for a type I transition, and 11 for a type

II transition. The reason 11 is assigned to a type II transition is that switching in different directions required changing the polarity of the charge stored in the coupling capacitance, hence, consuming about twice the amount of charge required for a type I transition. The codeword 11, instead of 10, helps to make a decision on data inversion using a majority voter, because the majority voter outputs high when at least eight input lines are high out of fifteen inputs. The majority voter can be implemented by using either full-adder circuitry or resistors and a voltage comparator. The control signal inv can be transmitted to the receiver using extra bus lines or extra transfer cycles. One problem of additional bus lines for control is the area overhead that may not be allowed due to physical constraints. In some cases, widening the space between signal bus lines can reduce the coupling effects more effectively than introducing extra control lines, because the coupling capacitance is inversely proportional to net space. Temporal redundancy is an alternative using extra clock cycles to transfer control signals.

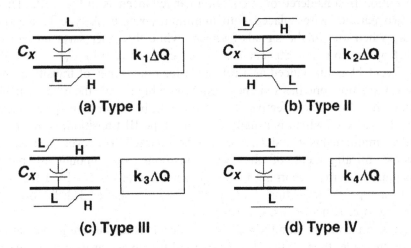

Figure 2-10. Transition types: (a) Single line switching; (b) both lines switching in opposite direction; (c) both lines switching in the same direction; (d) no switching.

2.4.2 Circuit techniques

One way to effectively increase noise immunity is to increase the switching threshold voltage V_{th} of the gate. V_{th} is defined as the input voltage at which the output changes state. Increasing the V_{th}, on the other hand, has an adverse effect on the performance such as speed and power consumption that are the prime features of dynamic circuits.

2.4.2.1 Gated-Vdd

Gated-Vdd is used to decrease the leakage power[30]. The key idea for gated-Vdd is to introduce an extra transistor in the supply voltage (Vdd) or the ground (Gnd) path. The extra transistor is turned on in the used section and off in the unused gated section. It maintains the performance and advantages of low power supply and threshold voltages while reducing leakage and leakage energy dissipation. The fundamental reason for the reduction in leakage is the stacking effect of self reverse-biasing series-connected transistors. However, it impacts the switching speed due to a non-zero voltage drop across the gated-Vdd transistor between the supply rails and the "virtual Gnd" for NMOS gated Vdd, Fig. 2-11, or the "virtual Vdd" for PMOS gated-Vdd.

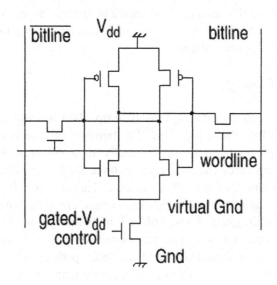

Figure 2-11. NMOS gated-vdd.

2.4.2.2 Dual threshold voltage

Dual threshold technique[31] is used to reduce leakage power by assigning high threshold voltage to some transistors in non-critical paths, and using low threshold transistors in critical paths. In order to achieve the best leakage power saving under target performance constraints, an algorithm is presented

for selecting and assigning an optimal high threshold voltage. Results show
that dual threshold technique is good for power reduction.

2.4.2.3 Dynamic threshold voltage

DTMOS (Dynamic Threshold MOS)[32] is a scheme that allows for a self-adjusting threshold voltage. By tying the gate to the body, the threshold
voltage decreases as the gate voltage increases, and vice versa. In this
manner, a higher zero-bias threshold can be used to reduce the leakage
current. In a speed-adaptive threshold-voltage CMOS (SA-Vt CMOS)
circuit[33], the substrate bias is controlled so that delay in the circuit stays
constant. Distributions of device speeds are squeezed under fast-operation
conditions. With a ring oscillator using 0.25-mm CMOS devices as a test
circuit, it was found that the worst-case operating frequency was improved
from 20 MHz to 55 MHz, and the fluctuation of the operating frequency was
suppressed from 44 % to 15 %, while the supply-voltage variation was under
0.1 V with a 1.8 V supply voltage.

2.4.2.4 C4 Flip-Chip

C4 Flip-Chip is used to manage the IR drop. IBM researchers developed
C4 technology in the 1960s, Fig. 2-12. The bonding process is characterized
by the soldering of silicon devices directly to a substrate (organic, for
example). The chip faces the substrate, as opposed to wire bonding, hence
the name flip-chip bonding. The salient features of this packaging
methodology are as follows: 1) Solder bumps are distributed on metal
terminals on the chip itself. These solder bumps are typically composed of
97% lead and 3% tin. The substrate has identically placed metal pads on its
surface. 2) The chip is turned over and the metal pads are aligned to solder
bumps; metal reflow is used to form connectivity between the substrate and
chip.

The advantages of C4 technology are numerous and important, as
follows:

- increased I/O density – C4 bumps may be placed over the entire area of
  the chip (called area array) rather than simply the periphery;
- self-aligning process step – due to surface tension;
- reduced die size for previously pad limited designs;
- reduced simultaneous switching noise due to smaller inductance of
  bumps compared to wire leads;
- better thermal properties as the backside of the wafer is now available for
  heat sinking;

- much better power distribution capabilities as circuits in the middle of the die can now access Vdd/Gnd directly;
- low cost and high throughput (all connections for one chip are made simultaneously in C4 as opposed to one-by-one in wirebonding);
- shorter wire lengths and fewer global wires ease wiring requirements.

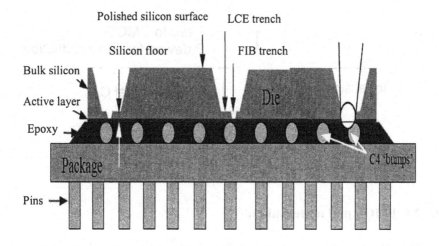

Figure 2-12. C4 Flip-Chip.

The main drawbacks to C4 at this time are that the use of lead in the solder bumps leads to the emission of alpha particles which can lead to circuit failure in sensitive circuits such as DRAMs. However, restricting the placement of solder bumps over these sensitive areas can minimize this effect. Research is ongoing to find alternate materials for solder bumps, as well. Also, the use of C4 packaging allows designers to do many different things in the floorplanning and routing stages of a design. Commercial tools for place-and-route, etc. are predicated on the use of peripheral wirebonding for I/O pads. New tools need to be in place for designers to take full advantage of flip-chip's advantages.

2.4.2.5 Pseudo CMOS

Pseudo CMOS, Fig. 2-13 provides the logic capability of domino and the noise robustness of CMOS.

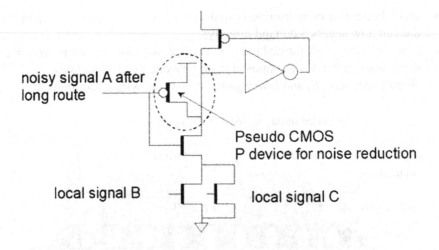

Figure 2-13. Pseudo CMOS.

2.4.2.6 PMOS pull up technique

The PMOS pull up technique, Fig. 2-14, utilizes a pull-up device to increase the source potential of the NMOS network thereby increasing the transistor threshold voltage V_t and V_{st} during the evaluate phase[34]. This technique suffers from large static power dissipation.

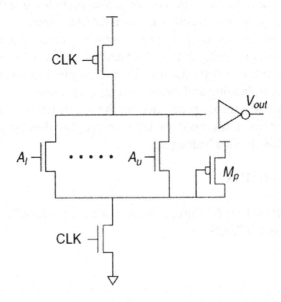

Figure 2-14. The PMOS pull up technique

2.4.2.7 CMOS inverter technique

The CMOS inverter technique utilizes a PMOS transistor for each input[35], (Fig. 2-15), thereby adjusting V_{st} to equal that of a static circuit. This technique cannot be used for dynamic NOR-type circuits, as certain logic combinations will short V_{dd} to GND.

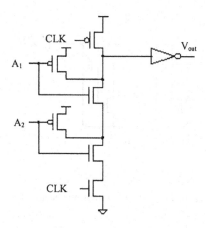

Figure 2-15. The CMOS inverter technique.

2.4.2.8 The Mirror technique

The mirror technique, Fig. 2-16, utilizes two identical NMOS evaluation networks and one additional NMOS transistor M_1 to pull up the source node of the upper NMOS network to $V_{dd} - V_t$ during the precharge phase[36] thereby increasing V_{st}. The mirror technique guarantees zero DC power dissipation, but a speed penalty is incurred if the transistors are not resized.

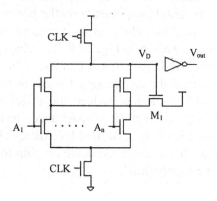

Figure 2-16. The mirror technique[36]

2.4.2.9 The twin-transistor

The twin-transistor technique, Fig. 2-17, utilizes an extra transistor for every transistor in the pull-down network in order to pull up the source potential[3738]. The twin transistor technique consumes no DC power.

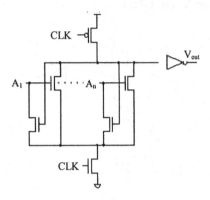

Figure 2-17. The twin transistor technique.

2.4.3 Architectural techniques

A high level description of a modified scheme for powering up/down resources is presented[39]. The scheme presented reduces the instantaneous current drawn form the supply when the resource is turned on and off and can be applied at different levels of the design abstraction. It is based on dividing the available resources in the system into smaller parts and then switching each part individually on and off at separate times instead of doing it instantaneously. This time separation between the on and off switching of the parts results in reduced sudden current pulses delivered from the supply and, hence, reduces the glitches appearing on the power and ground lines. A block diagram of the modified clock gating technique where the resource is divided into four parts is shown in Fig. 2-18. The clock signal is fed to each part through a separate AND gate that is controlled by an enable signal that is delayed with respect to the enable signal used for the previous resource. Another technique that is used to reduce the effect of the instantaneous current pulses is the decoupling capacitance technique. Decoupling capacitors have been effectively used to reduce dI/dt noise on printed circuit boards. This technique has become available on-chip to reduce the effect of inductive noise on the supply rails[40].

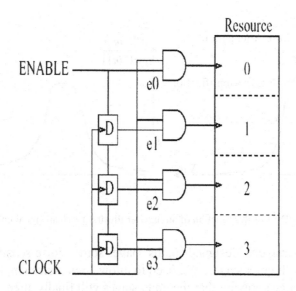

Figure 2-18. Using an enable signal to reduce inductive noise.

2.5 NOISE ANALYSIS ALGORITHMS

A specific level of noise is unavoidable in digital circuits. The question is to decide when it causes function failure. Unity gain is a criterion to decide when a circuit is considered unstable

2.5.1 Small signal unity gain failure criteria

Traditional analysis of noise margins relied on the small signal unity gain failure criteria[9]. For a small change in input noise to a circuit biased at an operating point, the resultant change in output noise is measured, Fig. 2-19. If $|d\ (Output)/\ d\ (Input)| > 1$ then the circuit is considered unstable. Unity gain is a good design metric but is neither necessary nor sufficient for noise immunity.

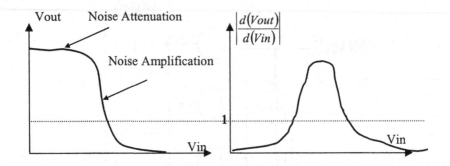

Figure 2-19. DC transfer function of an inverter illustrating small signal unity gain.

Most aggressively designed paths have some noise-sensitive stages interspersed with quiet stages. We need to allow some noise amplification in the sensitive stage knowing that the quiet stages will finally attenuate it.

2.5.2 Case study: Intel failure criteria

In this section, we list the steps and guidelines followed by Intel in determining a circuit failure.
- Break the circuit into circuit stages.
- Track the noise propagation across these stages by AC circuit simulation.
- Measure if any circuit stage failed due to
  - Injected noise combined with
  - Noise propagated from previous stages
- Noise can propagate across any number of stages eliminating the need for any unity gain budgeting.
- Combination of noise sources and simultaneous noise on multiple inputs should be considered.
- The peak noise in the event of two simultaneous couplers on a line is larger than the sum of these two events.
- The simultaneous occurrence of different parallel inputs has to be considered.
- New Transistor level (symbolic circuit) simulation should be found, since SPICE is very slow.
- Solving the differential equation symbolically in a piecewise linear manner can decrease the CPU time needed with good accuracy.

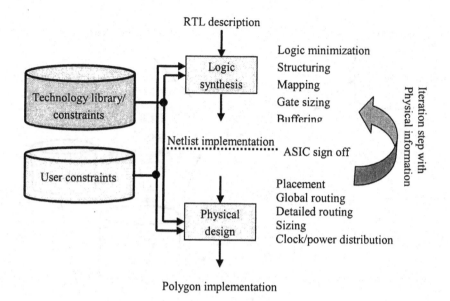

Figure 2-20. Traditional ASIC design flow.

The traditional design flow in Fig. 2-20 contains a separation between the logic synthesis step and the physical design step. For designs with aggressive performance goals, it was found that several iterations between synthesis and physical design are required to converge to a desired implementation[39].

As a result, design teams have begun to bring more of the backend design flow in-house, and the handoff to the semiconductor-vendor occurs only at the end. This approach is shown in Fig. 2-21 and is known as the customer-owned tooling (COT) approach.

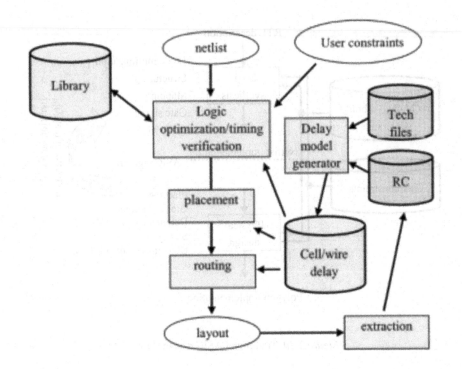

Figure 2-21. Today's high performance logical/physical flow.

Chapter 3

INTERCONNECT NOISE ANALYSIS AND OPTIMIZATION TECHNIQUES

The current trend of technology scaling is predicted to continue. Feature sizes will continue to shrink and clock frequencies to increase. Shrinking feature size implies not only shorter gate lengths, but also decreasing interconnect pitch and device threshold voltages[41]. The total wire area capacitance will decrease due to the reduction of the areas of a minimum-width wire. However, resistance is increasing faster despite efforts not to scale metal thickness. Moreover, the crosscoupling capacitance between neighboring wires will increase due to the decrease in spacing between wires. The expectation for the Die sizes is to remain constant in spite of the feature size shrinking. This is due to the integration of more functionality on a single chip. Practical efforts to control RC delays through the use of low-resistivity metal (copper), low-dielectric-constant insulators, and wide, thick wiring will require future interconnection analysis to consider inductance and inductive coupling. We present some interconnect models and categorize them. We also present some interconnect noise minimization techniques that are required in the nanometer technologies.

3.1. SILICON TECHNOLOGY

As stated before nanometer technologies require the adoption of new models and new optimization techniques that are capable of handling the new challenges. The difference between the older silicon technology and the current deep submicron technology[42] is shown in Fig. 3-1 and Fig. 3-2 respectively.

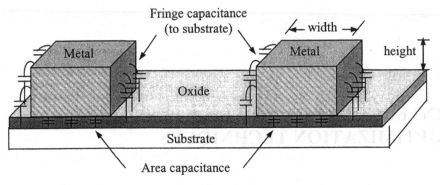

Figure 3-1. Older silicon technology

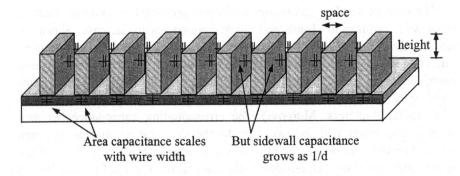

Figure 3-2. Deep submicron technology

The function of interconnects or wiring system is to distribute utility signals, such as clocks, and other signals, and to provide power/ground to and among the various circuits/systems functions on the chip. Current leading-edge logic processors have 6–7 levels of high-density interconnect, and current leading-edge memory has three levels[5]. There are three types of wiring to distribute the clock and signal functions (local, intermediate, and global). Local wiring, consisting of very thin lines, connects gates and transistors within an execution unit or a functional block (such as embedded logic, cache memory, address adder) on the chip. Local wires usually span a few gates and occupy first, and sometimes, second metal layers. Intermediate wiring provides clock and signal distribution within a functional block with typical lengths up to 3–4 mm. Intermediate wires are wider and taller than local wires to provide lower resistance signal/clock paths. Global wiring provides clock and signal distribution between the functional blocks, and delivers power/ground to all functions on a chip. The delay of local and global wiring in future generations is shown in Fig 3-3. Repeaters can be incorporated to mitigate the delay in global wiring, but they consume power and chip area.

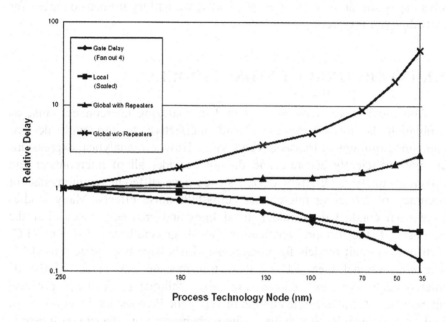

Figure 3-3. Delay for local and global wiring versus feature size

In the long term, new design or technology solutions (such as coplanar wave guides, free space RF, optical interconnect) will be needed to overcome the performance limitations of traditional interconnect[5]. Inductive effects will also become increasingly important as frequency of operation increases, and additional metal patterns or ground planes may be required for inductive shielding. As supply voltage is scaled or reduced, crosstalk has become an issue for all clock and signal wiring levels; the short-term solution adopted by industry is the use of thinner metallization to lower line-to-line capacitance. This approach is more effective for the lower resistivity copper metallization, where reduced aspect ratios (A/R) can be achieved with less sacrifice in resistance as compared with aluminum metallization. The 2001 Roadmap reflects this design trend by featuring reduced aspect ratios (as an alternative means of reducing capacitance) and less aggressive scaling of dielectric constant[43].

With technology scaling, noise is a problem affecting all types of designs from custom microprocessors to standard-cell ASICs. A noise analysis solution must be capable of analyzing tens of millions of transistors, considering both circuit and interconnect noise, and evaluating the distinct noise tolerances of each node in the circuit.

In this chapter, we survey new trends for interconnect synthesis and optimization techniques. We emphasize interconnect delay and noise models, techniques and algorithms to optimize them and comments on advantages and drawbacks of each. Finally, we will try to focus on areas for extending our research.

3.2. INTERCONNECT NOISE MODELS

Analytical expressions are preferred in analyzing interconnect noise as simulation is always expensive and ineffective to use with designs containing millions of transistors and wires. However, analytical expressions are not sufficiently accurate and do not consider all of interconnect and driver parameters. Different design stages have different requirements for accuracy of modeling interconnect crosstalk noise effects. Many studies have been conducted to model metal lines and crosstalk effects. For the purpose of the timing verification, ideal ground-based RC or RLC-distributed-circuit models for interconnect lines have been widely used[44,45]. The conventional ideal RC or RLC transmission line model of the IC interconnects without considering the following detailed physical phenomena are not accurate enough to verify the Pico-second level timing of high performance VLSI circuits[46]. These phenomena are the silicon substrate effect, the skin effect, and the proximity effect.

Most of the research efforts have been focused on developing formulas for the peak noise pulse amplitude. The pulse width of the crosstalk noise and the peak noise occurring time, which are of similar importance for circuit performance as the peak amplitude, should also be considered. Digital gates can filter out noise pulses with high amplitude provided that the noise pulse width is sufficiently narrow. Dynamic noise-immunity metrics, such as the noise immunity curve (NIC), are required[47]. Interconnect analytical models could be categorized into two categories: lumped and distributed models.

3.2.1 Lumped interconnect models

In the lumped model, the total capacitance and resistance values are used. A simple coupling circuit structure with one victim and one aggressor is shown in Fig. 3-4.

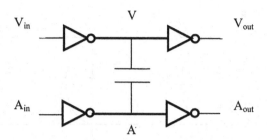

Figure 3-4. A simple coupling circuit structure

This is not the practical case as we always have more than one aggressor coupled to the victim line. The lumped capacitance model for this circuit is shown in Fig. 3-5. In this model, each pulling resistance, R_v or R_a, is composed of the line resistance and the driver resistance. The load capacitances C_a and C_v, consist of the line capacitance and the gate capacitance of the load driven by the line. C_m is the coupling capacitance between the two wires. Most of the crosstalk-noise-modeling papers published in the earlier years belong to this category[48-51]. In deep submicron technology, lumped models are no longer capable of satisfying the accuracy requirements and the consideration of some parameters like the coupling location.

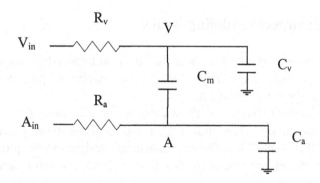

Figure 3-5. Capacitive coupling model for the circuit in Fig. 3-4

3.2.2 Distributed interconnect model

Distributed coupled RC (L) tree structure models become necessary even for the early design stages. Many papers addressed this problem[52-54]. The distributed π, 3π, 4π models have been used[55-57], respectively. We describe the π model in Fig. 3-6.

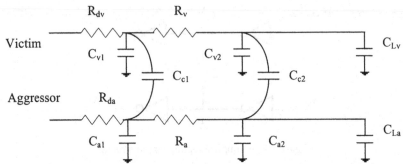

Figure 3-6. Interconnect distributed π model for the configuration in Fig. 3-4

In this figure, C_{Lv} and C_{La} are the victim and the aggressor load capacitances respectively. R_{dv}, R_{da} are the victim and the aggressor drivers modeled by resistances. C_{c1}, C_{c2} are the coupling capacitance between the two wires distributed to the beginning and the end of the wires. C_{v1}, C_{v2}, C_{a1}, C_{a2} are the distributed area capacitances for the victim and aggressor wires, respectively. In the 4π model[57], the users stated that their model is a complete analytical crosstalk noise model which incorporates all physical properties including victim and aggressor drivers, distributed RC characteristics of interconnects and coupling locations in both victim and aggressor lines.

3.2.3 Interconnect modeling issues

There are some issues that need to be considered while modeling the crosstalk noise such as the input signal shape, number of pins in a net, net topology, and the driver modeling[58,59].

The non-linear behavior of the victim driver gate during the transition should be captured[59]. The industrial noise analysis tool, ClariNet, has considered this issue. Results on industrial designs were presented to demonstrate the effectiveness of this tool with an 8% error compared to SPICE simulations.

It has been emphasized that the exact functional form that is used to estimate the capacitance and the delay are not important. The only requirement that the delay model must satisfy is that an increase [decrease] in the coupling capacitance should be translated into an increase [reduction] in the delay of a net[58].

Expressions for crosstalk amplitude and pulse width in resistive, coactively coupled lines should hold for nets with an arbitrary number of pins and of arbitrary topology under any specified input excitation[52]. It is claimed that the magnitude of inductive coupling is small in the presence of good ground return paths close to signal lines[52]. They used their expression in formulations aimed at reducing crosstalk. These methods include transistor sizing, wire ordering, wire width optimization and wire spacing. They failed to get success in some of these formulations. In wire ordering[48], they propose a new heuristic since they did not report results for this work.

The characteristics that lead someone to prefer one model among others include speed and accuracy. Depending on the design stage, different accuracies and speed are needed.

3.3. INTERCONNECT NOISE MINIMIZATION TECHNIQUES

Interconnect structures techniques, which are changing the interconnect topology, wire sizing, spacing, and buffer sizing and locations have been proved to be effective in reducing interconnect noise. These techniques come with the overhead of more chip area or more power consumption. Some other techniques like net ordering can prevent crosscoupling noise, but it needs pre-timing analysis for all nets and the determination of the sensitivity of nets to each other.

3.3.1 Buffer insertion

For long interconnects, wire sizing or spacing, explained in the next subsections, are not sufficient to limit the interconnect delay. So, buffer insertion is used to trade-off the active device area for reduction of interconnect delays. However, the additional inserted buffers increase the overall gate delay. Thus, there is an optimal number of repeaters that should be inserted into an RC line to minimize the overall propagation delay. This trend[60] is qualitatively illustrated in Fig. 3-7. During the routing of global interconnects, macroblocks form useful routing regions which allow wires to go through but forbid buffers to be inserted. They give restrictions on buffer locations. The buffer location restrictions have been taken into consideration and the simultaneous maze routing and buffer insertion problem have been solved in polynomial time[61].

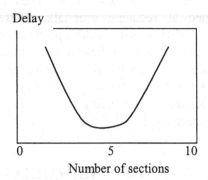

Figure 3-7. Relationship between the number of sections of an RC line and the total propagation delay

3.3.2 Wire sizing

It is known that proper wire sizing can effectively reduce the interconnect delay, especially in deep submicron or nanometer designs when the wire resistance becomes significant. An optimal wire-sizing algorithm was developed[62,63] for a single source RC interconnect tree to minimize the sum of weighted delays from the source to timing-critical sinks under the Elmore delay model. An efficient approach to perform global interconnects sizing and spacing (GISS) for multiple nets to minimize interconnect delays with consideration of coupling capacitance, in addition to area and fringing capacitances[64]. In that paper, they propose an asymmetric wire-sizing scheme where they may widen or narrow above and below the centerline of the original wire asymmetrically. The optimal wire sizing and spacing problem for a single net with fixed surrounding wire segments can be solved by adapting the bottom-up dynamic programming (DP)-based buffer insertion and wire-sizing algorithm[65].

3.3.3 Wire spacing

Wire spacing can decrease the crosscapacitance value effectively[66]. For example, The post-layout spacing heuristic[67] proposed a post-layout graph-based spacing algorithm. The framework spaces out wires after detailed routing has been finished. The crosstalk effect is based on crosstalk voltage glitches, as they do not consider the delay. The crosstalk effect from driver to sink is simply the superposition of all glitches of wires along the path. The potential disadvantage of this greedy operation is the poor utilization of space resources around a timing critical wire.

The graph-based optimizer preroutes wires on the global routing grids incrementally in two stages -- net order assignment and space relaxation[68]. The timing delay of each critical path is calculated taking into account interconnect coupling capacitance. The objective is to reduce the delays of critical nets with negative timing slack values by adding extra wire spacing.

3.3.4 Shield insertion

Shield insertion is another technique for decreasing crosstalk. It simply adds a power or a ground line between already existing wires. These power lines act like shields and can isolate between wires[69].

The existing net ordering formulations to minimize noise are no longer valid with the presence of inductive noise, and shield insertion is needed to minimize inductive noise[70]. Two simultaneous shield insertion and net ordering (SINO) problems are formulated: the optimal SINO/NF problem to find a min-area SINO solution that is free of capacitive and inductive noise, and the optimal SINO/NB problem to find a min-area SINO solution that is free of capacitive noise and is under the given inductive noise bound. They claim that this work is the first work that presents an in-depth study on the simultaneous shield insertion and net ordering problem to minimize both capacitive and inductive noise. The drawbacks of this work are the consideration for having the same wire lengths, and the consideration that the appropriate sensitivity matrix is given a priori. This matrix needs an overhead to be built, and they did not mention the complexity or how to build this matrix. They developed four approximate algorithms for solving the SINO/NB problem: greedy-based shield insertion (SI) algorithm, net ordering for minimizing C_x noise followed by SI algorithm (NO+SI algorithm), graph-coloring based SINO algorithm (SINO/GC algorithm), and simulated annealing based SINO algorithm (SINO/SA algorithm). They solve the SINO/NF problem by using the SINO algorithms and setting the noise bound to zero. The SINO/SA always performs the best for any given setting. This work has been extended by the same authors to consider RLC model[71].

3.3.5 Network ordering

It has been shown that signal ordering can effectively reduce the crosstalk noise on interconnects. In other words, arranging signals in an order that does not let a wire fight against its neighbor, making a transition in the same directions, can reduce the crosstalk noise. For example, two interconnect layout design methodologies for minimizing the "cross-coupling effect" in the design of full-custom datapath have been proposed[72].

These are the control signal ordering scheme and the track assignment algorithm. Fig. 3-8 shows how the control signal for a multiplexer can be ordered to decrease the crosstalk effect.

An evolutionary programming approach is used in track assignment to minimize the total number of tracks used. They propose an *"evolutionary programming-based track assignment (EPTA)"* algorithm considering both length and switching activity. It was shown that the solution quality of EP is quite good and the converging speed is very fast compared with SA and GA. More than anything else, EP does not require an annoying genetic operator design, which makes the implementation of EP very easy.

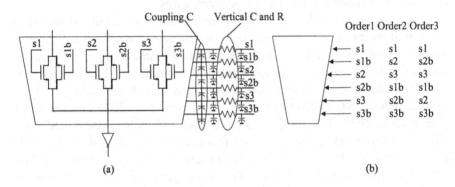

Figure 3-8. (a) A multiplexer example showing the signal coupling effects and (b) Three different orderings[72]

3.4. INTERCONNECT NOISE IN EARLY DESIGN STAGES

In the current design flow, interconnect optimizations are mainly used in post layout stages[73]. As the global interconnects are largely determined by floorplanning, it becomes critical for floorplanning engines to be able to handle efficient interconnect planning and optimizations, so that the overall timing and design convergence can be achieved. The overall framework for floorplanning and interconnect planning[73] is shown in Fig. 3-9. Due to the inherent complexity of the Interconnect Driven Floor Planning (IDFP) problem, they use multi-stage, adaptive cost functions within IDFP to gradually consider more interconnect optimization, planning, and/or global routing features. Using the adaptive cost functions gives them flexibility to tune for different design objectives, and to trade off between performance and run time. They use simulated annealing algorithm to achieve this flow. They divide the simulated annealing temperature region into four different stages from high to low temperatures. From Stage 1 to Stage 4, they use progressively more accurate interconnect performance and routability measurements as the temperature goes down. In Fig. 3-9 A, TL, ML, D, CJ1, CJ2 are the parameters used in the weighted cost function; and they mean area, total wire length, maximum length, maximum net delay, congestion estimation, and the average tile boundary congestion respectively. These terms are having adaptive relative importance in the cost function. This work adopted a net-based delay model, which is not practical.

The primary goal of some work was to examine global interconnect effects to determine if there are any significant road backs which will prevent National Technology Roadmap for Semiconductors (NTRS) performance expectations from being met[74]. Their analysis indicates that due to global RC delays as well as time-of-flight considerations, the global clock will necessarily be slower than the achievable local clock frequency. Also they have found that crosstalk at the global level will not be as significant as at the local level due to the use of large repeaters – their capacitance will dampen the effects of coupling capacitance. A wiring hierarchy which complements the modular design methodology is proposed[75]. This modular methodology proposes the use of 50,000 to 100,000 gate modules of logic to eliminate the impact of interconnect at the local level. These modules are arranged together in isochronous (or locally-clocked) regions, which run at a higher clock speed than the global clock. These isochronous come together to form the entire design.

A multi-layer gridless detailed routing system for deep submicrometer physical designs is presented[76]. Their detailed routing system uses a hybrid approach consisting of two parts: 1) an efficient variable-width, variable-spacing detailed routing engine and 2) a wire-planning algorithm providing high-level guidance as well as ripup and reroute capabilities. They use a nonuniform grid graph, which has been proved to guarantee a gridless connection of the minimum cost in multilayer variable-width and variable-spacing routing problem. They suggested further improving their wire-planning algorithm and fine tuning of ripup and rerouting algorithm. An integer linear programming (ILP) formulation that yields tight bounds on the quality of the achievable wire-packing solution has been formulated[77]. The input to Crosstalk-aware wire-packing comprises a set of wires, a set of variable tracks (per layer), and a crosstalk graph XG that determines forbidden wire adjacencies. The output is a legal assignment such that no forbidden adjacencies and no electrical shorts exist, as in Fig. 3-10. The key technical insight is to model all constraints (both geometric and crosstalk conflicts) as cliques in an appropriate conflict graph; these cliques can be extracted quickly from the interval structure of the slice.

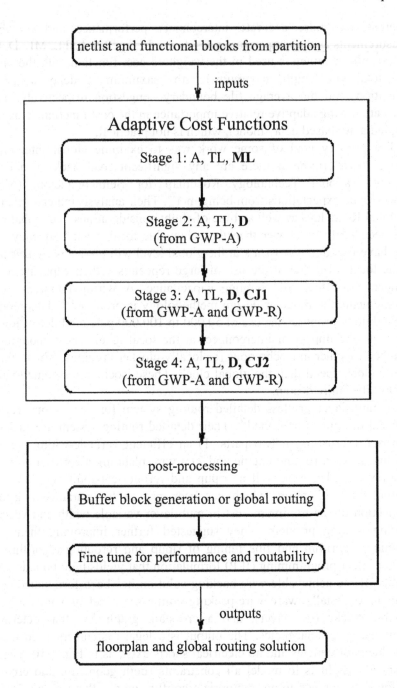

Figure 3-9. The overall flow of interconnect-driven floorplanning and global wiring planning and optimization[73]

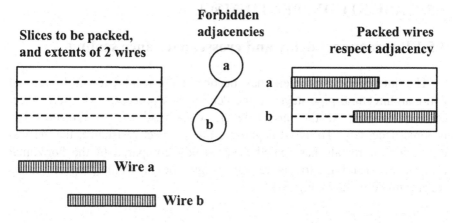

Figure 3-10. Wire Packing instance[77].

Methods to reduce interconnect delay and noise caused by coupling have been developed[78]. The coupling-free routing (CFR) takes a set of nets and tries to find a one-bend couple-free routing for a subset of nets. A routed net must not be coupled with any other routed net. They define coupling as a Boolean variable which is true when the coupling is greater than some threshold. They develop an exact algorithm for the CFR decision problem via a transformation to 2-satisfiability. This algorithm runs in linear time. They also present the implication graph which models the dependencies associated with CFR. They also develop an algorithm for the maximum coupling-free layout (MAX-CFL) problem. Given a set of nets, the MAX-CFL is defined as finding a subset of nets that are coupling-free routable. The subset should have maximum size and/or criticality. They call this algorithm "implication algorithm." They present the coupling capacitance between two wires i and j as follows:

$$C_c(i, j) = \frac{f_{ij} \cdot l_{ij}}{d_{ij}} \frac{1}{1 - \dfrac{w_i + w_j}{2d_{ij}}}$$

where w_i and w_j are the sizes of wires i and j ($w_i, w_j > 0$), f_{ij} is the unit length fringing capacitance between wires i and j, l_{ij} is the overlap length of wires i and j and d_{ij} is the distance from the center line of wire i to the center of wire j. Since the nets are routed at most one bend, they have minimum wire length. In addition, coupling-free routing minimizes the coupling of the routed nets.

It has been shown that the delay of a wire of length l increases at the rate of $O(l^2)$ without wire sizing, $O(l\sqrt{l})$ with optimal wire sizing and linearly with proper buffer insertion[79]. The criticality function can easily be changed to incorporate some other functions.

3.5. CASE STUDY: PENTIUM® 4

3.5.1 Interconnect delay and crosscapacitance scaling

The interconnect problem has become significant enough to require entire architectural pipe stages in the Pentium® 4 processor for interconnect communication[9]. At the circuit level, widespread use of repeaters has become necessary. To avoid degrading interconnect resistance, the vertical dimension of metals has scaled very weakly compared to the horizontal dimension, leading to extremely high height/width aspect ratios (approximately 2-2.2), Fig. 3-11.

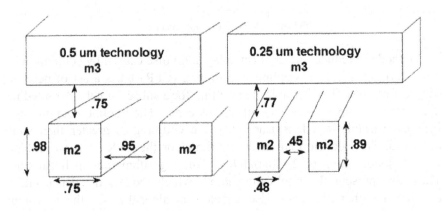

Figure 3-11. Wire aspect ratio scaling with technology [9]

Nowadays, the crosscoupling capacitance between parallel neighboring wires, which can get routed together for long distances, are becoming of great importance compared to the wires self-capacitances. This can either lead to a large increase in delay, coupling noise, min delay, or power consumption problems, depending on the switching direction of neighboring wires. Avoiding these delay and noise problems would involve drastically increased wire spacing or extensive shielding. Further, studies on both the Pentium® III and Pentium® 4 processor floor plans have clearly shown that they tend to be interconnect-limited for Die area, which increases the penalty for spacing and shielding. Thus, there is a fundamental design tradeoff between a simple, robust, wiring solution employing extensive spacing and shielding vs. an aggressive solution employing short wiring with only judicious shielding leading to high density. The latter requires sophisticated CAD tools, has more risks, but ultimately is much more optimal for a high-volume product. It was therefore the choice for the Intel Pentium® 4 processor[9].

3.5.2 Wire and repeater design methodology for the PENTIUM® 4 processor

Delay, noise, slope limits, and gate oxide wear out were all considered when drafting the guidelines for the wire and repeater methodology. Notable features were an increased emphasis on noise robustness and "pushed process" considerations for delay (repeater distance guidelines were made shorter than optimal for delay with the existing process, in anticipation of end-of-life process trending when transistors speed up a lot compared to wires). Repeater sizing, rather than best delay optimization for non-coupled wires, was picked to be optimal for noise rejection, for equal rise and fall delays, and for better delay in the presence of coupling. Stringent limitations were put on maximum sizing of repeaters, especially in buses, to reduce power supply collapse caused by a simultaneously switching bank of repeaters. The methodology and tools allowed Intel developers to use both inverting and non-inverting repeaters. Simple length-based design rules were provided for repeaters, and further optimization was possible through internally developed proprietary tools: NoisePad, ROSES, and Visualizer (net routing and timing) analysis. The extensive use of dedicated repeater blocks is evident in the Pentium® 4 processor floorplan. Further, the net length comparison in Fig. 3-12 shows that although the Pentium® 4 processor is a much larger chip, there are very few long nets in it compared to previous-generation chips such as the Pentium® III processor. This is even more notable given that the Pentium® 4 processor has more than twice as many full-chip nets as the Pentium® III processor and has architecturally bigger blocks. If we compare the M5 wire segments of the Pentium® III, and Pentium® 4 processors, we note that 90% of the M5 wire segments of the Pentium® 4 processor are shorter than 2000 microns while the same percentage of Pentium® III processor wires are 3500 microns long. These short wires are a key to enabling high-frequency operation.

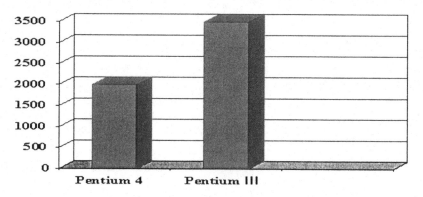

Figure 3-12. M5 length comparison of global wires for different processors[9]

Chapter 4

CROSSTALK NOISE ANALYSIS IN ULTRA
DEEP SUBMICROMETER TECHNOLOGIES

As discussed before, with scaling down of feature sizes to ultra deep submicrometer (UDSM) dimensions and high clock frequencies, signal integrity becomes one of the most vulnerable problems in deep submicron circuits[80]. Serious on-chip electrical problems are being encountered in these high-speed UDSM circuits. These problems include signal distortion along coupled interconnect lines, voltage variations in the power supply distribution, charge sharing, and substrate coupling[81]. Crosstalk noise, which is known as coupling noise, imposes three side effects on digital design. It can affect timing, causing a delay failure; it can increase the power consumption due to glitches, and it can cause functional failure. Coupling capacitance between neighboring nets is a dominant component in today's deep submicron designs as taller and narrower wires are laid out closer to each other[5]. It frequently has been reported in the literature that an IC interconnect delay dominates a critical path delay[82]. Since the timing budget and noise margins will become much tighter with the technology advance, the problem due to the IC interconnect lines will be more apparent.

In this chapter, crosstalk noise, as well as impact of coupling on aggressor delay is analyzed. The pulse width of the crosstalk noise, which is of similar importance for circuit performance as the peak amplitude, is also measured. We consider parameters like spacing between wires, wire length, coupling length, load capacitance, rise time of the inputs, place of overlap (near driver or receiver side), frequency, shields, direction of the signals, wire width for both the aggressors and the victim wires. Also, we consider parameters like driver strength as several recent studies considered the simultaneous device and interconnect sizing.

An important observation about the results obtained in this work is that we were able to get the effects of some interconnect parameters like the place of crosscoupling, the signal direction between two parallel wires, and the operating frequency.

4.1 ANALYTICAL EXPRESSIONS

Analytical expressions are preferred for performance evaluation because simulation is always expensive and ineffective in use with modern designs containing millions of transistors and wires. Different design stages have different requirements for accuracy of modeling crosstalk noise effects.

For the purpose of the timing verification, ideal ground-based RC or RLC-distributed-circuit models for interconnect lines have been widely used[44, 45]. The conventional ideal RC or RLC transmission line model of the IC interconnects, without considering the following detailed physical phenomena, are not accurate enough to verify the Pico-second level timing of high performance VLSI circuits[46]. These phenomena are the silicon substrate effect, the skin effect, and the proximity effect.

4.2 TRANSMISSION LINE MODEL

4.2.1 General transmission line

A transmission line in SPICE consists of one or more conductors and a ground layer or a ground line. The ground layer is called reference layer or layer stack[83]. A transmission line is shown in Fig. 4-1 with its distributed lumped parameters. HSPICE uses the telegrapher's equation given below to solve voltages and currents in transmission lines.

$$-\partial v(z,\omega)/\partial z = (R(\omega) + j\omega L)i(z,\omega)$$

$$-\partial i(z,\omega)/\partial z = (G(\omega) + j\omega C)v(z,\omega)$$

The R, L, G, and C are matrices and v and i are vectors.

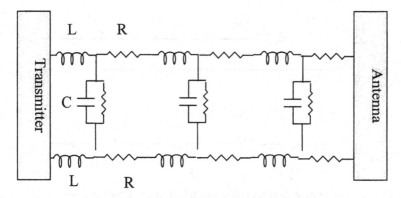

Figure 4-1. Transmission line model

This model is used assuming TEM model of wave propagation through the wire. At any instant of time the solution is similar to the static solution. But, conductors do not always display only TEM property[84]. Some conductors (especially in the presence of dielectrics) have longitudinal field components. That is why a quasi-TEM model is used. Resistance and conductance matrices R and G are not totally frequency independent due to skin effect and rotation of dipole under the alternating field. These are calculated using the equations shown below.

$$R(f) = R_o + \sqrt{f(1+j)}R_s$$

$$G(f) = G_o + \frac{f}{\sqrt{1+(\frac{f}{f_{gd}})^2}} G_d$$

R_s is the skin-effect matrix and G_d is the power loss due to dipole effect. R_o is the DC resistance and G_o is the conductance through the imperfect dielectric.

4.2.2 Transmission line with two conductors

The transmission line, considered here, is of three rectangular conductors. In order to consider the very general case for two parallel wires, which is being able to vary the total wire lengths and the cross coupling length separately, we use three transmission lines, Fig. 4-2.

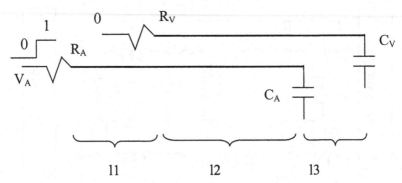

Figure 4-2. Three transmission line model, 2 conductors

The first transmission line is one conductor of length l1, the second transmission line consists of two conductors of length l2, and the third transmission line consists of three conductors of length l3. The aggressor length is l1+l2, the victim length is l2+l3, and the crosscoupling length between the two wires is l2. Using three transmission lines allowed us to vary l1, l2, and l3 separately, and to be able to accurately notice the effects of each of the wire length and the crosscoupling length.

4.2.3 Transmission line with three conductors

In real circuits, we rarely have victim nets, which are coupled to single aggressors only. The final output is expressed as the summation of the individual outputs produced by switching one aggressor at a time[50,54]. However, the crosstalk noise computed using such an approach cannot serve as an upper bound or as a lower bound for the actual noise in the original coupling circuit[85]. Therefore, the existence of multiple aggressors has to be explicitly accounted for. In this section we show the second model used for the transmission line simulation. It is simply one transmission line with three conductors of the same length. It has been simulated to see the effect of having more than one aggressor at the same time making transitions in the same or opposite directions. Another reason this model was built is to see the effects of having the aggressor and the victim drivers at the same or opposite sides as shown in Fig. 4-3.

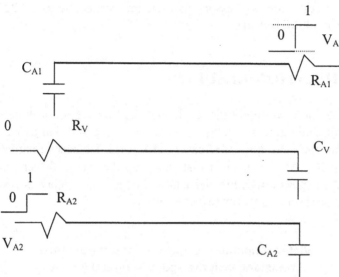

Figure 4-3. Coupling circuit structure

4.2.4 Transmission line with seven conductors

As inductance is a long range phenomena, the mutual inductance can be between a victim and its direct neighbor, its next to direct neighbors or further than this. When we consider that we have seven conductors with the one in the middle being the victim, the nearest two wires are the shields and the remaining four wires are the aggressors with two from each side.

4.2.5 Peak noise and delay parameters

In this section we list the circuit parameters that had been simulated to see the effect of each one on the peak noise amplitude, noise pulse width and the aggressor delay. Some parameters come from noise avoidance techniques[86] and others from the circuit itself. In summary, the parameters are spacing between the parallel wires, wire length, crosscoupling length, wire size, driver strength, load capacitance, coupling location, rise time, frequency, and signal and driver directions.

In the test circuits, the parameter ranges were as follows: R_{ag} (aggressor driver strength): 40-500Ω, R_v (victim driver strength); 40-500Ω, C_{lv}, C_{la} (load capacitances for aggressor and victim lines respectively); 10-100fF, l_v, l_a (victim and aggressor lengths respectively); 100-2000μm, l_c (coupling length) and coupling locations in aggressor and victim lines; 100μm-maximum length, t_r (rising time); 10-1000ps. Sp (spacing); 0.11-0.55μm. w_a, w_v (wire width for aggressor and victim respectively); 0.22-1.8μ, f (frequency); 500MHz-5GHz.

4.3 SIMULATION RESULTS

In this section, we report the results obtained for victim peak noise, noise pulse width, and aggressor delay. It is shown in Fig. 4-4 that as we increase the spacing, the peak noise decreases. The noise decreases relatively faster for a small R_{ag}. Also, it is clear that sizing up the victim driver (scaling R_v down) reduces peak noise. Fig. 4-5 shows the aggressor delay vs. spacing for different aggressor and victim driver sizes.

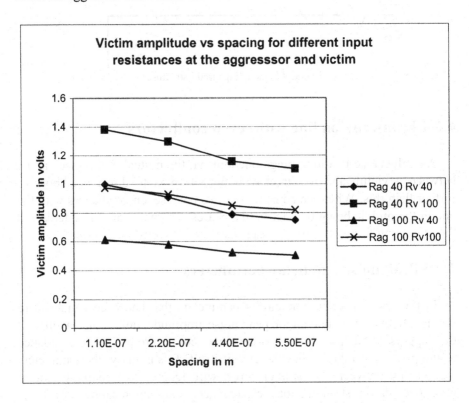

Figure 4-4. Victim amplitude vs. spacing (different Rs)

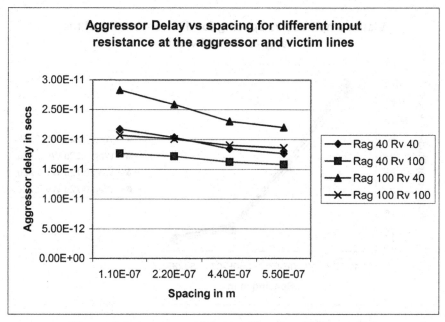

Figure 4-5. Aggressor delay vs. Spacing (different Rs)

As a wire's width increases, its resistance decreases and its ground capacitance increases. The aggressor and victim widths have approximately no effect on the noise amplitude, but we see how effective spacing is for a 1900μm coupling length (l_c=1900μm), Fig. 4-6. This result should not be generalized. The correct notice should state that the effect of wire sizing diminishes if the coupling location is close to victim driver. Our experiments show that wire sizing is very effective when coupling location is close to victim receiver.

For the effect of sizing on the aggressor delay we show the results in a table format, Table 4-1, as it is not clear through graphs.

Table 4-1. Aggressor delay vs. spacing

Spacing	Aggressor delay			
	W_a=22μm W_v=22μm	W_a=22μm W_v=72μm	W_a=72μm W_v=22μm	W_a=72μm W_v=72μm
1.10E-07	9.40E-12	9.34E-12	9.42E-12	9.54E-12
2.20E-07	8.16E-12	8.20E-12	8.12E-12	8.22E-12
4.40E-07	6.81E-12	6.81E-12	6.76E-12	6.81E-12
5.50E-07	6.37E-12	6.40E-12	6.34E-12	6.36E-12

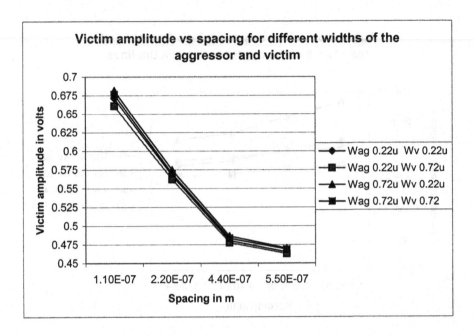

Figure 4-6. Victim amplitude vs. spacing

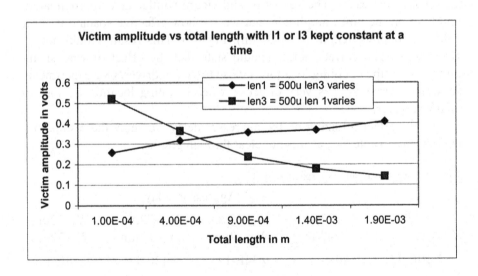

Figure 4-7. Noise amplitude vs. wire length

The results in Fig. 4-7 and Fig. 4-8 show the effect of the aggressor and victim wire lengths with a constant coupling length between them. The peak noise decreases as the aggressor length increases as long as the coupling length is kept constant. As seen in Fig. 4-8, the victim length has no effect on the aggressor wire delay.

Fig. 4-9 shows how the aggressor peak noise increases with the increase of the coupling length. This, of course, is due to the increase of the coupling capacitance. In this kind of experiment, we varied, in addition to the coupling length, both of l1, l3 to maintain constant total wire lengths leaving us with the coupling length as the only varying parameter.

Another important parameter that affects the noise especially in high-speed circuits is the input rise time. Fig. 4-9 and Fig. 4-10 show that increasing the rise time can decrease the noise amplitude scarifying some of the circuit speed.

Digital gates are inherently low-pass filter, and thus, can filter out noise pulses with high amplitude provided that the noise pulse width is sufficiently narrow. Dynamic noise-immunity metrics, such as the noise immunity curve (NIC), are required[47]. Fig. 4-11 and Fig. 4-12 show how the pulse width changes with different interconnect parameters. When we have a look at both Fig. 4-4 and Fig. 4-11, we see how effective spacing is to decrease both the noise pulse amplitude and noise pulse width.

For a given line length, the victim amplitude decreases with increase in shield width and it increases, with increase in frequency. However, this regularity was not observed for increasing line lengths, Fig. 4-13.

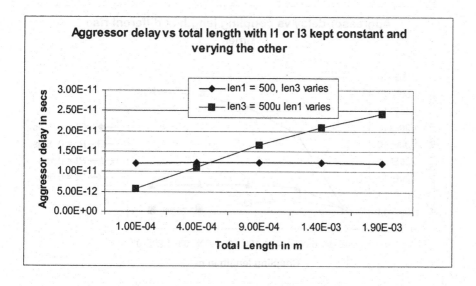

Figure 4-8. Aggressor delay vs. wire length

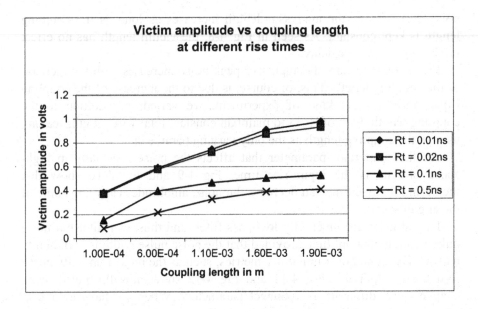

Figure 4-9. Victim noise amplitude vs. coupling length

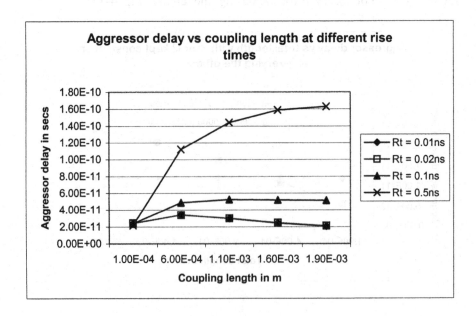

Figure 4-10. Aggressor delay vs. coupling length

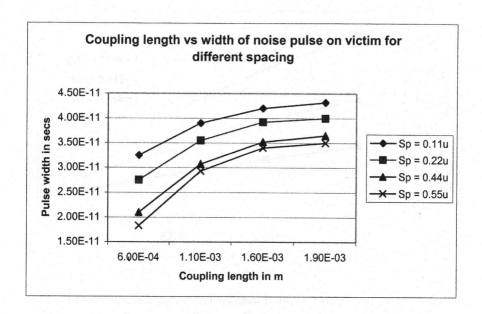

Figure 4-11. Width of noise pulse vs. coupling length for different spacing

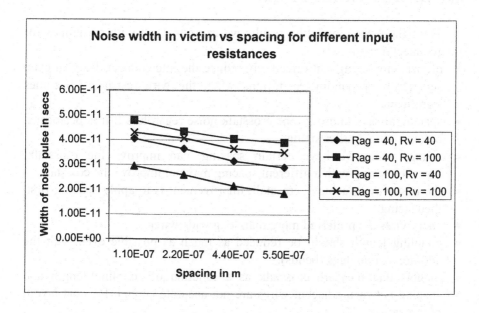

Figure 4-12. Width of noise pulse vs. spacing for different drivers sizing

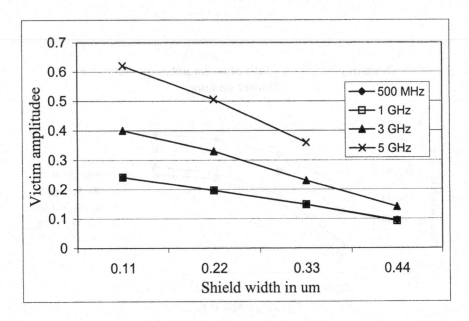

Figure 4-13. Victim amplitude vs shield width for different frequencies with line length = 2000μm

4.4 DESIGN GUIDELINES

Throughout the simulation results we can come out with guidelines for interconnect design:

- proper wire sizing can effectively reduce the interconnect delay in deep submicron or nanometer designs when the wire resistance becomes significant.
- spacing always improves the crosstalk noise regardless if it is performed before sizing or after sizing.
- sizing, while we have certain spacing, can improve the crosstalk; however, sizing with a different spacing may not improve the crosstalk.
- the best way to achieve crosstalk improvement is to apply spacing first, then sizing.
- short wires are preferred more than long wide wires.
- coupling length should be reduced as much as possible even with the additions of doglegs (bends).
- models that measure crosstalk as a function of coupling length and spacing between adjacent wires are not accurate enough for post layout optimization.
- place of overlap (near driver or sink side) is an important factor for crosstalk noise calculation; we should avoid coupling near sink side.

- increasing shield width can decrease inductive coupling noise, but increasing the number of shields used is more effective in decreasing this noise.
- noise pulse width is an important metric for measuring the noise; it is as important as noise peak amplitude.
- aggressor and victim driver strengths can affect the crosstalk noise; the higher the victim driver strength, the lower it is susceptible to noise.
- wire width is not that effective for crosstalk noise; it is a very important factor for delay.
- input rise time is an important factor that can affect the noise; the slower the input is the lower the noise generated.

4.5 SUMMARY

In order to verify functional failures or timing violations in high-speed circuits, we need to measure three metrics for the noise. These metrics are peak noise amplitude, noise pulse width, and peak noise occurring time. We have described the possible interconnect and driver parameters that affect crosstalk noise. We showed the effect of each of these parameters: spacing, length, coupling length, etc. on the noise metrics. We also showed the aggressor delay sensitivity to these parameters. Throughout the results obtained using SPICE, we were able to analyze and evaluate various noise avoidance techniques like increasing the spacing between two adjacent lines and driver sizing. We reported a set of guidelines for interconnect design and optimization.

Chapter 5

MINIMUM AREA SHIELD INSERTION FOR INDUCTIVE NOISE REDUCTION

With high clock frequencies, faster transistor rise/fall time, long signal wires, and the use of wider wires and Cu material interconnects, inductance of interconnect and the noise generated because of this inductance is becoming an important design metric in digital circuits. For a risk-free layout solution of a chip, capacitive and inductive noises should be considered at various routing process stages. An efficient technique to reduce the inductive noise of on-chip interconnects is to insert shields among signal wires. These shields are quiet wires that are either connected to the ground or to the power supply. A formulation and efficient solution for the min-area shield insertion problem to satisfy given explicit noise bounds in multiple coupled nets is provided. The proposed algorithm determines the locations and number of shields needed to satisfy certain noise constraints. The noise model used can handle different wire widths, different spacing among wires, and different wire lengths. The model is also aware of the skin effect in high frequency ranges. Experimental results show that the proposed approach gives the minimum number of shields to satisfy the noise constraints and uses less runtime than the best alternative approach.

5.1 INDUCTIVE COUPLING

Feature sizes will continue to shrink to very deep submicrometer (VDSM) dimensions and clock frequencies will increase. Shrinking feature size implies not only shorter gate lengths but also decreasing interconnect pitch and device threshold voltages. Reduction in the top and bottom areas of a minimum-width wire means that total wire capacitance is decreasing.

Resistance, however, is increasing faster, despite efforts not to scale metal thickness. Practical efforts to control RC delays through the use of low-resistivity metal (copper), low-dielectric-constant insulators, and wide, thick wiring will require future interconnection analysis to consider inductance and inductive coupling.

Mutual inductance noise occurs when signal switching causes transient current to flow through the loop formed by the signal wire and current return path[9], thereby creating a changing magnetic field (see Fig. 5-1). This induces a voltage on a quiet line, which is in or near this loop. Mutual inductance is a long-range phenomenon, which means that the inductive coupling could affect adjacent and non-adjacent wires. High-speed switching and synchronous bus structures are making this noise very significant in current technologies. Because the analysis of inductive effects is highly dependent on layout and is quite complex, the Pentium 4 design approach was to design the problem out through rules rather than analyze arbitrary configurations[9].

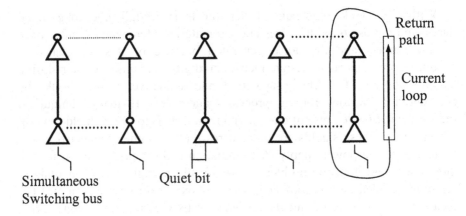

Figure 5-1. Mutual inductance noise from simultaneous switching on a wide bus

Wire sizing, spacing, and network ordering are some interconnect optimization techniques that are used to reduce crosscoupling capacitive noise. However, these techniques do not help to reduce inductive noise and may even worsen it. An efficient technique to decrease the inductive noise is to insert shields among signal wires. These shields are quiet wires that are either connected to the ground or to the power supply. These shields can be inserted in any routing layer in the layout. The shield insertion solution has the drawback of increasing the routing area, so we assume that we have spacing resources available for these shields. Also, shields that are inserted between coupled wires that are making transitions in the same directions at the same time may worsen the crosscoupling capacitive noise. This is due to the cancellation of the Miller effect.

Coupling inductance has been illustrated[87], where the coupling of an 18-bit bus is computed by SPICE simulations using RLC model for the wires. It has been shown that the noise as a percentage of Vdd can be decreased from 55% with 0 shields to 29% Vdd with 2 shields and further decreases to 13% Vdd with 5 shields. Increasing the shield width can decrease the noise amplitude[69]. However, the decrease in the noise amplitude, as shown in Fig. 4-13, was not that much compared to the insertion of another shield. Based on these experiments, it has been concluded that, in general, inserting more shields is more effective than increasing the shield width.

The formulation of simultaneous shield insertion and net ordering for capacitive and inductive coupling minimization has been studied before[70]. It has been proven that this problem is NP-hard. Different heuristic algorithms have been proposed and analyzed. A comparison between a noise free structure and a structure that can accept some noises to a certain bound is presented. The results are valid except that they have some major drawbacks in their initial assumptions. The algorithms assume that the sensitivity between all pairs of segments in the layout is known. A priori knowledge of these sensitivities requires timing analysis[88], which can represent a considerable overhead. Finding this sensitivity relationship also requires the existence of a detailed layout. One more comment about that work is the use of what they called the effective coupling model, which they did not validate in their work. The effective coupling model has a 15% error as was reported and does not consider the segment length or width. We expect this already high error percentage to increase when they consider wires with different lengths and widths. Some other recent interconnect optimization for RLC crosstalk reduction includes twisted bundle layout structure[89], and differential signaling[90].

We formulate the shield insertion problem and propose an efficient algorithm that determines the locations and number of shields needed to satisfy certain noise constraints for a min area. What is meant by min area is min number of inserted shields. Our algorithm supports any inductive noise model that is aware of the wire geometry and the frequency in use. The model we used can handle different wire widths and different wire lengths that are coupled to more than one segment. The model is also aware of the skin effect in the high frequency ranges. The algorithm consists of three phases. In the first phase of the algorithm, we determine the locations of the shields to be inserted without actually inserting them, yet taking the effect of each shield into consideration. During the second phase, we get the intersection of the shield locations and determine the min number of shields needed. In the last phase, we actually update the layout and insert the shields generated from phase two.

5.2 PROBLEM FORMULATION

5.2.1 Preliminaries

Throughout our work, we consider parallel coplanar interconnect structures with different wire lengths. This is in contrast to the previous work[70], in which the model and algorithms considerations are limited to interconnect structures with all wire segments having the same length. The proposed work can consider more than one routing layer at the same time, but only one layer layout is discussed for simplicity. Fig. 5-2 shows two signal lines that are between two power/ground (P/G) lines acting as the shields. A shield is simply a wire directly connected to a power/ground structure. In this figure, l is the wire length, t is the wire thickness, w is the wire width, and d is the distance between the two wires.

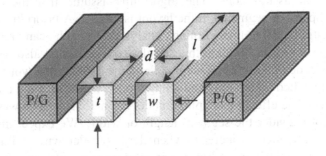

Figure 5-2. Interconnect structure

Because of proximity effects in high frequencies, a window enclosing the nets that couple to each other can be defined, which is bounded by nearest power/ground lines. Current returns of the nets are assumed to be through these power/ground lines[91]. We call the group of wires between adjacent P/G lines a block. We assume no crosstalk (coupling) between different blocks. This means that no signal in a block is sensitive to any segment in other blocks.

Horizontal shield insertion is considered as a case study. The vertical shield insertion can be performed similarly. We strongly suggest that horizontal and vertical shield insertion for two layer routing of a net should be performed simultaneously.

Let $N = \{n_i\}$ be a set of the nets in a chip, and let $H = \{h_j\}$ be a set of the horizontal interconnection segments. Each net n_i is built from horizontal segments h_{ij} and vertical segments. We assume that the most top and bottom segments are P/G lines. We consider the derived ITRS[5] 0.10μm technology (see Table 5-1). This table acts as a guide for the range of parameters used, and it can be varied. The maximum inductive noise that a net can withstand can vary from one net to another. For different noise constraints, we calculated the maximum inductance allowed[92]. We assume that the sensitivities among different nets are known, in spite of the overhead to generate these sensitivities.

Table 5-1. Interconnect specifications based on derived ITRS 0.10μm technology

Vdd	1.05V	Load capacitance	60fF
Frequency	3GHz	Wire width	1.0μm
Input rising time	33ps	Wire thickness	1.1μm
Driver resistance	150Ω	Wire spacing	0.8μm

5.2.2 Net sensitivity

If a switching signal on a net (aggressor) can cause a failure on another net (victim), these two nets are considered sensitive to each other. The aggressor may cause noise in more than one victim at the same time. The sensitivity between nets can be represented with a sensitivity matrix of size n x n (where n is the number of nets). By definition, the sensitivity matrix is symmetric since if one net is sensitive to the other, the reverse relation should be true. Fig. 5-3 shows an example of a sensitivity matrix for five nets. An entry of 1 in location (i,j) indicates that net_i and net_j are sensitive to each other. A 0-entry indicates that the two nets are not sensitive to each other.

In the experimental results section, we will explain the criterion we adopted for generating different matrices for a specified sensitivity rate.

$$
\begin{array}{c}
\quad\quad A\ B\ C\ D\ E \\
\begin{array}{c} A \\ B \\ C \\ D \\ E \end{array}
\left(\begin{array}{ccccc}
0 & 1 & 0 & 1 & 1 \\
1 & 0 & 0 & 0 & 0 \\
0 & 0 & 0 & 0 & 1 \\
1 & 0 & 0 & 0 & 0 \\
1 & 0 & 1 & 0 & 0
\end{array}\right)
\end{array}
$$

Figure 5-3. A sensitivity matrix for 5 nets

5.2.3 Noise modeling

For inductance calculations, we use the analytical formulae for self and mutual inductance estimation[91].

Equation (1) is for the self-inductance when $l >> (w + t)$ where l is the wire length, w and t are the width and thickness of the rectangular cross section, respectively. Equation (2) is for the mutual inductance of two parallel wires of distance d and equal length when $l>d$,

$$L_{self} = \frac{\mu_0}{2\pi}\left[l\ln\left(\frac{2l}{w+t}\right)+\frac{l}{2}+0.2235(w+t)-\mu_r l(0.25-X)\right] \quad (1)$$

X=0.4372x   if x<0.5
X=0.0578x+0.1897 if $0.5 \le x \ge 1$
X=0.25   if x>1

$$\chi = \frac{\delta}{0.2235(w+t)}, \text{ Where } \delta \text{ is the skin depth at a particular frequency}$$

$$M = \frac{\mu_0 l}{2\pi}\left[\ln\left(\frac{2l}{d}\right)-1+\frac{d}{l}\right] \quad (2)$$

To calculate the mutual inductance of parallel wires with unequal length, we use the previously derived equations[91]. The six different cases are shown in Fig. 5-4. We will mention the different cases of wire geometries used for the calculations and give one formula as an example.

The mutual inductance of a wire with the inducing wires is the algebraic sum of the mutual inductance of the separate elements of the inducing wires. Considering case 3, the mutual inductance can be calculated as shown in Equation (3),

$$M = \frac{1}{2}\left[\left(M_{m+p}+M_{m+q}\right)-\left(M_p+M_q\right)\right] \quad (3)$$

If a wire consists of several segments of which self-inductances are known, the whole wire's self-inductance does not equal to the sum of all the self-inductances of all the segments because of the existence of mutual inductances. Instead, all segments' self-inductances as well as mutual inductances between these segments are required to compute the whole wire's self-inductance. The self-inductance of a whole wire constructed by cascaded segments is

$$L_{self} = \sum_{i=1}^{N} l_i + \sum_{i=1}^{N} \sum_{j=i+1}^{N} 2k_{ji} M_{ij} \qquad (4)$$

where N is the number of segments, l_i is the self inductance of segment I, Mij is the mutual inductance between segment i and j of the whole wire. K_{ij} =0 when segment i and j are orthogonal; K_{ij} =1 when i and j have the same current direction; K_{ij} =-1 when i and j have opposite current direction.

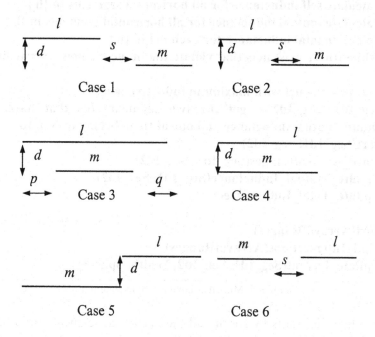

Figure 5-4. Six relative positions for mutual inductance

5.3 SHIELD INSERTION ALGORITHM

We present an inductive crosstalk optimization approach. Given as its input a feasible routing solution of the chip, risk tolerance bounds of nets, our approach produces a minimum area risk-free routing solution in which the layout is free of all inductive noise violations. We call it minimum area as it gives the minimum possible number of shields to be inserted. The overview of the algorithm is shown in Fig. 5-5.

Sort {hj} by its y-axis coordinates in an ascending order
Calculate self inductance for all horizontal segments in {h$_j$}
Calculate mutual inductance for all horizontal segments in {h$_j$}
Calculate total inductance for each net in {n$_i$}
While (there exist nets that violate the inductive noise constraints)
{
Net_Id = the net with maximum inductive noise
Seg_Id1, Seg_Id2 = get the two segments Ids that have the maximum mutual inductance and one of them belong to Net_Id
Sort(Seg_Id1, Seg_Id2)
Store_ArrayofRanges(Seg_Id1, Seg_Id2)
Update_Mutual_Inductance(Seg_Id1, Seg_Id2)
Update_Total_Inductance()
}
Sort(ArrayofRanges)
Find_Intersections(ArrayofRanges)
Update_Layout(Seg_Id1, Seg_Id2, Width, Spacing)

Figure 5-5. Min-area shield insertion algorithm

The algorithm starts by sorting all the horizontal segments from bottom to top and assigns segment numbers to them after that sort. The very bottom segment will have segment Id 1 and the very top segment will have segment Id m. We use the inductance calculation equations in Section 5.2 to calculate mutual inductance and self inductance for all segments and store them. Then, we calculate the total inductance for all nets. The algorithm keeps iterating to find nets that violate the inductive noise constraint and fixes them until no more nets that violate the constraints exist.

The main idea of the algorithm is that whenever a shield is needed to be inserted, we postpone this shield insertion until we identify all shield insertion positions. We store the positions (ranges) where we should have inserted this shield until we finish the iterations.

In each iteration we identify the net with maximum noise violations using the percentage of violation, (Net_Inductance – Inductance_Bound) /

Net_Inductance * 100, instead of the absolute noise number, Net_Inductance. For that net, we find the segment that has the largest mutual inductance with all other segments inside its block. The reader can find the block definition in Section 5.2. Now, we identified two segment Ids (range) where we need to insert a shield between them. We postpone the shield insertion until we identify all ranges. However, we consider the effect of the shield inserted immediately. This is done through the Update_Inductance functions. In the Update_Mutual_Inductance function, we set the mutual inductance between all segments above Seg_Id2 and all segments under Seg_Id1 to zero. This is done because we are sure that a shield is going to be inserted between these two segments and the segments above Seg_Id2 and under Seg_Id1 will be definitely in different blocks. The function Update_Total_Inductance is to subtract the mutual inductance from nets that we have set some of its segments mutual inductances to zero.

The second part of the algorithm is to get the intersection between all these ranges to minimize number of shields inserted. We first sort the array in which we stored ranges. This is done for implementation issues and then we get the ranges intersections in the Find_Intersections function. The function pseudo code is shown in Fig. 5-6.

We demonstrate our algorithm through a simple example. Consider the layout shown in Fig. 5-7. In this figure, only the horizontal segments are shown for simplicity. It is clear that we can have segments of different lengths and different spacing between them as stated before. The segments have segment Ids in ascending order from bottom to top. Throughout the iterations of the algorithm, we found some nets that violate the noise constraints, and we determined some ranges to insert shields in. Assume that the first part of the algorithm generated these ranges to insert shields in.

(4,11), (6,11), (6,10), (5,9), (4,8), (3,7), (3,5), (1,4)

Part 1 of the algorithm generated eight shields to be inserted between different segments. If we consider the first 2 ranges (4,11) and (6,11) which means we need a shield to be inserted between segment 4 and segment 11, and a shield to be inserted between segment 6 and segment 11. This maps to one shield between segments 6 and 11. The following is a summary for the rest of the ranges:

Considering (6,10) the resulting range is (6,10)
Considering (5,9) the resulting range is (6,9)
Considering (4,8) the resulting range is (6,8)
Considering (3,7) the resulting range is (6,7)
Considering (3,5) the resulting range are (6,7) and (3,5)
Considering (1,4) the resulting range are (6,7) and (3,4)

Therefore, we are left with only two shields that need to be inserted, one between segments 3 and 4, and the other one between segments 6 and 7.

Phase 3 actually updates the layout and inserts the shields. A flowchart is given in Fig. 5-8. This part, which is implemented in the Update_Layout function, accepts, as an input, the two segment Ids. Seg_Id1, Seg_Id2 to insert a shield between, the shield width, Width, and how much spacing between the shield and Seg_Id1, Spacing. It generates the updated layout with the inserted shields after modifying the horizontal and vertical segments for a two-layer layout. It is a very simple task to be extended to a multi-layer layout.

```
    trunkid1 = 0;
    trunkid2 = MaxInt;
    Initialize IntersectionResults to zeros;
    We start by having ArrayofRanges ordered in descending
order of trunkid1;
    For each rang element in ArrayofRanges array {
    trunkid1 = Max(ArrayofRanges[i].trunkid1,
IntersectionResults[noOfShields].trunkid1);
    trunkid2 = Min(ArrayofRanges[i].trunkid2,
IntersectionResults[noOfShields].trunkid2);
     if(trunkid1 < trunkid2)
     {
    IntersectionResults[noOfShields].trunkid1 = trunkid1;
    IntersectionResults[noOfShields].trunkid2 = trunkid2;
     }
    else
        {
          noOfShields++;
          IntersectionResults[noOfShields].trunkid1 =
ArrayofRanges[i].trunkid1;
          IntersectionResults[noOfShields].trunkid2 =
ArrayofRanges[i].trunkid2;
        }
     }
    noOfShields++;
```

Figure 5-6. Find_Intersections function pseudo code for the second phase

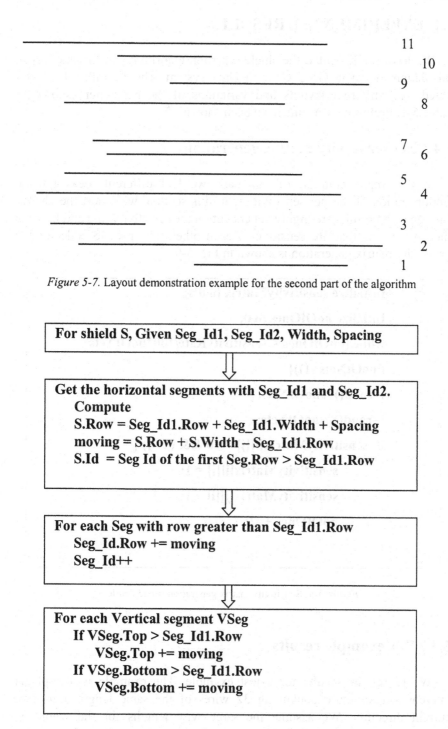

Figure 5-7. Layout demonstration example for the second part of the algorithm

For shield S, Given Seg_Id1, Seg_Id2, Width, Spacing

Get the horizontal segments with Seg_Id1 and Seg_Id2.
Compute
S.Row = Seg_Id1.Row + Seg_Id1.Width + Spacing
moving = S.Row + S.Width – Seg_Id1.Row
S.Id = Seg Id of the first Seg.Row > Seg_Id1.Row

For each Seg with row greater than Seg_Id1.Row
Seg_Id.Row += moving
Seg_Id++

For each Vertical segment VSeg
If VSeg.Top > Seg_Id1.Row
VSeg.Top += moving
If VSeg.Bottom > Seg_Id1.Row
VSeg.Bottom += moving

Figure 5-8. Flowchart to insert a shield and update layout

5.4 EXPERIMENTAL RESULTS

We have implemented the shield insertion algorithm and the inductance calculation model in GNU C++ on Unix system. The algorithm has been tested with different layouts and variations of the parameters shown in Table 5-1. Preliminary results have been shown[93,94].

5.4.1 Net sensitivity matrix generation

In the implementation of this part, we had different ideas for the randomization of the net sensitivities. In this section, we report the chosen one, since the results are highly dependent on the sensitivity rate and how we choose the nets that are sensitive to each other. The pseudo code for the sensitivity matrix generation is shown in Fig. 5-9.

```
Initialize sensitivityMatrix to 0's;

Intialize noOfOnes to 0;

while (noOfOnes < sensitivityRate/100*noOfNets*

( noOfNets -1)){

i=rand()%noOfNets;

j=rand()%noOfNets;

if (sensitivityMatrix[i][j] == 0 && i != j) {

        sensitivityMatrix[i][j] = 1;

        sensitivityMatrix[j][i] = 1;

        noOfOnes +=2;

}
```

Figure 5-9. Sensitivity matrix generation pseudo code

5.4.2 Bus example results

We report the results for a test example. The example is a coplanar interconnect structure containing 32 wires of the same length and same current direction. We assume the same wire lengths for the whole 32 segments at a time and fixed spacing between segments (edge-to-edge spacing = 0.8µm). All wires are 1.1µm thick, 1µm wide, as shown in

Fig. 5-10. We consider two lengths of 1400μm and 2000μm, and four sensitivity rates of 25%, 30%, 50%, and 60%. A demonstrative example after inserting three shields, as determined by the algorithm, is shown in Fig. 5-11. The dashed lines are the three shields inserted. The results are shown in Table 5-2 through Table 5-5. The fraction part comes from the average for several runs. In our experiments, we get different number of shields from different runs while having the same sensitivity rate. The reason for that is the random assignment of the signals that are sensitive to each other. In the implementation of the algorithm, we only considered wire lengths in which inductance effects are significant[60].

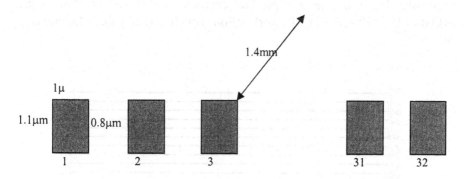

Figure 5-10. A coplanar interconnect structure containing 32 wires of the same length

Table 5-2. Number of shields needed by the algorithm for sensitivity rate = 25%

noise bound	length = 1400μm	length = 2000μm
0.15V	4.46	5.36
0.20V	4.26	5.27

Table 5-3. Number of shields needed by the algorithm for sensitivity rate = 30%

noise bound	length = 1400μm		length = 2000μm	
	Proposed	Compared	Proposed	Compared
0.15V	4.73	4.85	5.40	5.40
0.20V	4.32	4.40	5.29	5.05

Table 5-4. Number of shields needed by the algorithm for sensitivity rate = 50%

noise bound	length = 1400μm	length = 2000μm
0.15V	7.65	7.93
0.20V	6.89	7.67

Table 5-5. Number of shields needed by the algorithm for sensitivity rate = 60%

noise bound	length = 1400µm		length = 2000µm	
	Proposed	Compared	Proposed	Compared
0.15V	7.69	7.60	7.98	7.85
0.20V	6.88	6.55	7.71	7.45

As can be seen in Table 5-3 and Table 5-5, the number of shields generated by the proposed algorithm and the compared one[70] are very close. The difference ranges between –2.5% to 5.0% for the reported example.

The algorithm has been proven O (N^2) where N is number of horizontal segments. The run time for the bus example is < 1 sec on "Sun Ultra SPARC-III+" 900 MHz CPU workstation, which is better than the previous results[70].

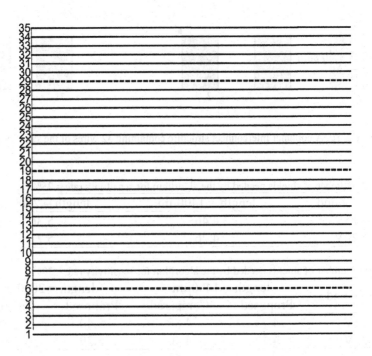

Figure 5-11. A demonstrative example of 32-wire bus after inserting three shields

5.5 COMPLEXITY ANALYSIS

The complexity of the proposed shield insertion algorithm is computed as shown in Fig. 5-12.

Total Complexity = Complexity of pre-processing + Complexity of algorithm part I + Complexity of algorithm part II

Complexity of pre-processing = Sort segments + Calculate self inductance for segments + Calculate mutual inductance between segments + Calculate total inductance for each net =

$O(N\log N) + O(N) + O(N^2) + O(N) = O(N^2)$

Complexity of algorithm part I = get net with max inductive noise + get 2 segments with max mutual inductance in that net + update mutual inductance + update total inductance =

$O(N) + O(N^2) + O(N^2) + O(N^2) = O(N^2)$

Complexity of algorithm part II = Complexity of sorting array of ranges + Complexity of find range intersections =

$O(N\log N) + O(N) = O(N\log N)$

Hence, total complexity of the algorithm is $O(N^2) + O(N^2) + O(N\log N) = O(N^2)$

Figure 5-12. Complexity analysis of the proposed shield insertion algorithm

5.6 SUMMARY

We have formulated the min-area shield insertion problem to satisfy inductive noise constraints. We have presented an efficient algorithm that minimizes the number of shields needed to satisfy noise constraints. The algorithm complexity is found to be of $O(N^2)$, where N is the number of horizontal segments in the layout. The main idea of the algorithm was to postpone the actual shield insertion until all required shields are identified, and then find the intersection of all the ranges to minimize the number of shields (area) needed to satisfy the noise constraints. The run time to optimize a 32-segment bus has been reported to be < 1 sec with no noise violations in the result. The algorithm has shown to be flexible and give good results for different geometries, different noise bounds and different sensitivity rates.

Chapter 6

SPACING ALGORITHMS FOR CROSSTALK NOISE REDUCTION

We have shown in the previous chapters that the features size will keep shrinking, the wire geometric ratio becomes larger, and wires will have less spacing to neighbor wires. Due to these factors, crosscoupling capacitance will increase affecting the wire delay and signal integrity on that wire. A very useful technique to reduce the crosstalk noise is to increase spacing between wires, which we call re-spacing. However, this re-spacing technique may consume resources like chip area. In this chapter we discuss several available algorithms that use this concept. Some of these algorithms consider spacing alone and others consider spacing and other techniques, like sizing, at the same time.

6.1 SIMULTANEOUS WIRE SIZING AND WIRE SPACING IN POST-LAYOUT

The wire sizing and spacing problem has been studied[94]. The Elmore delay model was used and it accounts for coupling and ground capacitance in this model. They construct section constraint graph in each routing region and use the graph to guide segment sizing and spacing. They do not minimize the interconnect delay, but they expect that the interconnect delay is not too far away from its given bound. They construct a cost function that accounts for the trade-off in dealing with timing and area and formulate the problem as a constraint-optimization problem.

In that work, the authors did not deal with noise on interconnect, they only optimized for the wire delay. Also, they did not provide the criteria to divide the channel into these regions. They considered only uniform sizing.

They did not account for the delay from vertical segments and crosscoupling between vertical segments. They mentioned that optimizing the vertical segments will easily cause Design Rule Check (DRC) problems, but they did not consider their effect in their cost function. The algorithm starts from an initial section constraint graph with minimum size and spacing for all wires and tries to minimize the number of edges in this graph. So, this algorithm cannot keep the benefits of previous efforts maintained during routing to select suitable sizing and spacing for wires, and the whole idea of the algorithm will be destroyed if it tries to account for previous sizes and spaces. This leads to the conclusion that it is not suitable for tools with incremental routing, which is of great importance for designers in the industry. One last comment about this work is about the step size for increasing wire width or spacing. They chose it to be uniform and did not justify this choice and how large should it be. Also, we think that this algorithm cannot skip a local minima as it always increases the size and space and can never go back to check other choices.

6.2 POST GLOBAL ROUTING CROSSTALK SYNTHESIS

A post global routing crosstalk optimization approach is proposed[95]. It estimates and reduces crosstalk risk at the global routing level. They use Integer Linear Programming to minimize the crosstalk risk among nets. They address crosstalk optimization in global routing. The approach at the global routing level is region-based, which estimates the crosstalk risk for each routing region on the chip as a whole and reduces the risk by adjusting nets' routes globally among routing regions on the chip. Instead of generating a specific risk-free final solution for each region, they produce a global routing solution of the chip in which a risk-free final solution of each region exists. Their approach is graph-based. They assume that both the sensitivity information and risk tolerance bounds of nets can be extracted using temporal and functional analysis or specified by user.

The disadvantage of this approach is very clear. They use Integer Linear Programming, which does not scale well for large problem size. It may take days without being able to get a solution. The assumption of the availability of the sensitivity information is not correct. This process is very time consuming and is not that accurate on the global level. They also assume that a net occupies a whole horizontal or vertical track, which is not practical and has a major effect on their technique to handle the problem. One last comment about this work is that they consider only shielding technique to remove noise. Shielding is considered a technique that consumes area and power compared to the spacing technique[96].

6.3 TIMING- AND CROSSTALK-DRIVEN AREA ROUTING

Processing the crosstalk and timing constraints by ordering nets and tuning wire spacing in a quantitative way is presented[67]. That approach enjoys a larger optimization solution space than the previous approaches whose solution space is highly limited by routed geometric constraints. In that work, a graph-based optimizer preroutes wires on the global routing grids incrementally in two stages – net order assignment and space relaxation. The timing delay of each critical path is calculated taking into account interconnect coupling capacitance. The objective is to reduce the delays of critical nets with negative timing slack values by adding extra wire spacing. They handle this spacing technique by assigning the positions of cross points along the boundaries of routing regions.

6.4 A SPACING ALGORITHM FOR PERFORMANCE ENHANCEMENT AND CROSSTALK REDUCTION

A post-layout graph-based spacing algorithm is proposed[66]. The framework spaces out wires after detailed routing has been finished. The crosstalk effect is based on crosstalk voltage glitches, as they do not consider the delay. The crosstalk effect from driver to sink is simply the superposition of all glitches of wires along the path. The cost function is, therefore, more for the measure of logic fault hazards rather than timing fault hazards. However, it more or less overestimates the crosstalk glitches because it does not consider the resistance–capacitance (*RC*) filter effect on a path from driver to sink and directly sums up all the unweighted glitches on a path. They utilize a graph-based algorithm, which selects the node that has the worse coupling and increases the edge weight between them.

The potential disadvantage of this greedy operation is the poor utilization of space resources around a timing critical wire. Also, the formulation as an iterative parameterized linear programming (IPLP), as mentioned by the authors, cannot handle the coupling between the vertical segments of the wires (instead, the vertical coupled lengths are incorporated into the objective function). This approach is also slow because of the large number of linear programs that it needs to solve.

6.5 A POST PROCESSING ALGORITHM FOR CROSSTALK-DRIVEN WIRE PERTURBATION

Post-processing algorithms can use accurate crosstalk measurements to respace the wires, thus improving crosstalk even in routings produced by crosstalk aware routers[97]. In that work, they minimize the peak crosstalk in the nets of a gridless channel. They divide the spacing range into intervals where the crosstalk is linear in each of these intervals avoiding solving nonlinear equations. They consider crosstalk in both horizontal and vertical segments of a net at the same time. They assume that the total crosstalk is proportional to the sum of the capacitive couplings of (or noise contributions) in each segment of the net with each of its neighbors. The disadvantage of this work is that they do not have an explicit formula for the noise, and they do not consider the net delay at all. The following example shows that capacitive coupling can be decreased and at the same time the delay may increase.

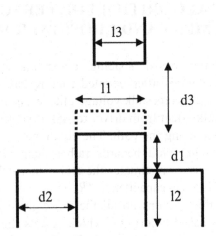

Figure 6-1. Capacitive coupling and delay behavior

In Fig. 6-1, Assume that l1 = 30, d1 = 3, l3 = 2, d3 = 2
According to the formula used to calculate the crosstalk
Crosstalk = l1 C / d1 + 2 l2 C / d2 + l3 C / d3 = l1 C + 2 l2 C / d2
If we moved the l1 segment up until d1 = 4, d3 will be 1 and the new crosstalk calculation will be 9.5 C + 2 l2 C / d2. the l2 term is the same in both cases and, as we see, the crosstalk in the second case is less than the one in the first case. However, the vertical segments lengths have been increased which may lead to delay increase and worsen the performance instead of improving it using this model.

Chapter 7

POST LAYOUT INTERCONNECT OPTIMIZATION FOR CROSSCOUPLING NOISE REDUCTION

As VLSI technology continues to move towards nanometer, the coupling capacitance between adjacent nets and the noise generated due to this coupling are becoming increasingly major concerns in high-speed designs[2]. The coupling capacitance increase in nanometer technology is due to the increase in the aspect ratio and the decreasing spacing between adjacent nets. The noise generated because of the high coupling capacitance can be a function failure, a timing failure (increased delay), and/or the consumption of more power. In the previous chapters, we presented how spacing is effective in reducing capacitive crosstalk noise and surveyed some of the previous work that utilized this interconnect parameter for noise reduction. In this chapter, we present a proposed re-spacing framework to reduce crosstalk noise. We explain the noise model used and how it is extended to handle simultaneous multiple crosscoupling. The algorithm and the model used can handle the effect of re-spacing among signals in one layer and the other connected segments in another layer. We will present the details for a 2-layer layout for simplicity. The extension to multi-layer layout is very simple and straightforward. We present a simple iterative algorithm that processes critical paths in decreasing order of crosstalk noise. The experimental results for the tested benchmarks are promising and encourage applying this framework for industrial layouts.

7.1 MOTIVATIONS

For existing routers, even those which are crosstalk-aware, wires that are placed into their final layout position must use approximations for their couplings with neighbors that have not yet been put in place[98]. Rip-up and reroute can resolve the routing-resource-violations and improve the crosstalk among nets. However, significant improvement still can be achieved by post-processing algorithms. Although post-processing algorithms suffer the draw back of having less flexibility in moving the wires around, they can use accurate models and measurements of the crosstalk in the nets. So, our post-processing layout re-spacing can improve the crosstalk existing in a routed layout. We use the same available routing resources to redistribute the crosstalk among nets to get a more reliable layout.

In post processing, we need to have accurate measurements for the noise as this is considered the last step in the design flow before final tests. It is not enough to measure crosstalk between nets by considering only the coupling capacitance[98]. We have shown a counter failure example for that work. Many of the parameters should be considered in the calculation of the crosstalk noise. By crosstalk noise, we mean not only the noise amplitude, but also the noise pulse width as well[99].

The framework presented here is different from the previous work[98], introduced in Section 6.5, in many aspects – the measure of the crosstalk effect, the routing topology; i.e., channel routing or general area global routing, and the selection of segments to be replaced by the algorithm. What we mean by replacing a wire segment is to put it in a new position by moving it up/down (left/right) for a horizontal (vertical) segment. The measure of crosstalk[98] is based on crosscoupling capacitance using the formula $\dfrac{Cl^{\eta}}{d^{\gamma}}$ where C is the unit crosstalk, l is the coupling length, d is the separation between the two wires, and $\gamma > 0$ and $\eta \geq 1$. In our work, we measure an accurate peak noise amplitude and noise pulse width taking into account the location of coupling, the driver strength, and the load capacitance. The measured crosstalk over estimates the crosstalk glitches because it does not consider the resistance-capacitance (RC) filter effect on a path from the driver to a sink and directly add the unweighted crosscoupling in a net.Channel routing and the replacing of only the horizontal segments, which they call trunks are the only considerations[98]. In our work, we consider a general area global routing and allow any of the horizontal or vertical segments to be respaced and replaced in a new position as long as we do not violate the design rules. For the selection of a segment to be replaced, we have implemented two different mechanisms. In the first one, we select the segment, which has the largest effect on the crosstalk noise. This means that for two segments with the same crosscoupling (same overlap length, same spacing, and the same crosscoupling capacitance constant), we will consider the segment near the sink node to be replaced first. This is because it has the greater effect on decreasing noise. The second mechanism is a very simple one, in which we consider segments to be replaced in the following order. We start from the driver side and consider all the segments along the path until we reach the sink side. The reason for this simple methodology is that we will be sure that the resistance value from the driver until the segment under consideration will not be changed later by replacing further segments. This will ensure an accurate calculation for the crosstalk noise in each segment. The last difference that we want to stress is that in contrast to the previous work[98], in which they consider the crosscoupling for the whole net, we consider the crosstalk noise in a net for each path from the driver to any of its sinks, which form many paths.

7.2 PROBLEM FORMULATION

Assume that we have a pre-routed 2-layer layout with horizontal and vertical segments. As mentioned before, the extension to multi-layer layout is easy and trivial. In the noise model we adopted, we consider all the circuit and interconnect parameters that affect the calculation of the crosstalk noise. Examples of these parameters are the driver strength, driver rise time, load capacitance, and wire widths.

We assume that we have a routing of n nets $N_1, N_2, ..., N_n$. The routing of these nets consists of horizontal and vertical segments. For net N_i, ($i=1, ..., n$). its horizontal segments are $H_1^{(i)}, ..., H_{h_i}^{(i)}$, where h_i is number of horizontal segments in net N_i, and its vertical segments are $V_1^{(i)}, ..., V_{h_i}^{(i)}$, where v_i is number of vertical segments in net N_i.

Let the total number of horizontal and vertical segments be h and v, respectively, ($h = \sum_{i=1}^{n} h_i$, $v = \sum_{i=1}^{n} v_i$), so that we can also enumerate the horizontal and vertical segments as $H_1, ..., H_h$ and $V_1, ..., V_v$, respectively. An example layout is shown in Fig. 7-1.

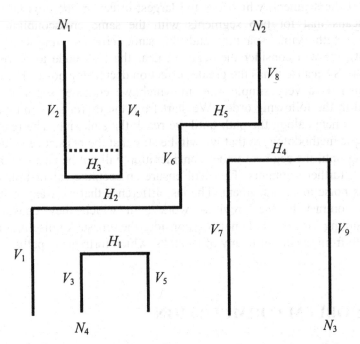

Figure 7-1. Example layout with 4 nets, 5 horizontal segments, and 9 vertical segments

We try to minimize the maximum noise in each net by redistributing the spacing among segments taking into account the effect of segments in the other layer. For example, considering Fig. 7-1, moving horizontal segment H_3 up to the position of the dotted line may decrease the coupling noise between segments H_3 and H_2. Also, it may decrease the noise between segments V_4 and V_6 and the total noise in net N_1 may decrease. To accept this spacing redistribution, we must be sure that all of the other affected nets, N_1. s neighbors do not get worse than the net under consideration, N_1. Thus, our problem can be stated as

$$\text{Minimize } \max_{i=1}^{i=n}\{noise_i\}\tag{1}$$

7.3 NOISE MODELING

We adopt an already existing noise model, extend it for handling multiple crosscoupling, and provide the equations for the variation in capacitance resistance due to wires moving by our respacing algorithm. The noise model used is the improved 2-π crosstalk model for noise constrained interconnect optimization[50,100].

For a victim net with some aggressor nearby, as shown in Fig. 7-2(a), let the aggressor voltage pulse at the coupling location be a saturated ramp input with transition time (i.e., slew) being t_r, and the interconnect length of the victim net before the coupling, at the coupling and after the coupling be L_s, L_c and L_e, respectively.

The 2-π type reduced RC model is generated as shown in Fig 7-2(b) to compute the crosstalk noise at the receiver. It is called 2-π model because the victim net is modeled as two π-type RC circuits, one before the coupling and one after the coupling. The victim driver is modeled by effective resistance R_d. Other RC parameters C_x, C_1, R_s, C_2, R_e, and C_L are computed from the geometric information from Fig. 7-2(a) in the following manner. The coupling node (node 2) is set to be the center of the coupling portion of the victim net, i.e., $L_s + Lc /2$ from the source. Let the upstream and downstream interconnect resistance/capacitance at Node 2 be R_s/C_s and R_e/C_e, respectively. Then, capacitance values are set to be $C_1 = C_s/2$, $C_2 = (C_s + C_e)/2$ and $C_L = Ce/2 + C_1$.

The closed form peak noise amplitude and the effective noise width are given below. The effective noise width is defined as the length of time interval that noise spike voltage v is larger or equal to a given certain threshold voltage level v_t, Fig. 7-3.

$$v_{max} = \frac{t_x}{t_r}(1 - e^{-t_r/t_v}) \qquad (2)$$

$$t_{width} = t_r + t_v \ln\left[\frac{1 - e^{-2t_r/t_v}}{1 - e^{-t_r/t_v}}\right] \qquad (3)$$

where,

$$t_x = (R_d + R_s)C_x \qquad (4)$$

$$t_v = (R_d + R_s)(C_x + C_2 + C_L) + (R_e C_L + R_d C_1) \qquad (5)$$

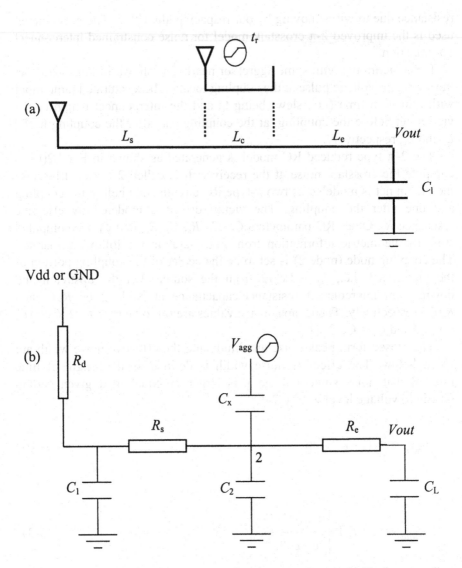

Figure 7-2. (a) The layout of a victim net and an aggressor above it. (b) The 2-π crosstalk noise model

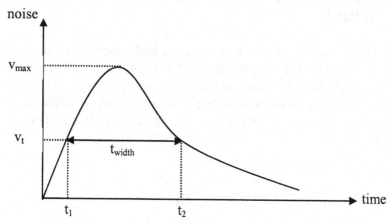

Figure 7-3. Illustration of the noise width

For a victim net in general RC tree structures, the model can be easily extended. To compute the crosstalk noise at a certain sink, the 2-πmodel is built as shown in Fig 7-4. The lumped capacitance at each branch on the path from source to sink; i.e., C_{b1}, ... C_{bi}, is incorporated in the following weighted manner:

- if a branch B_i is between the source and the coupling center, let its distance to the source be $\alpha(L_s+L_c/2)$. Then $(1-\alpha)C_{bi}$ goes to C_1 and αC_{bi} goes to C_2.
- if a branch B_i is between the sink and the coupling center, let its distance to the sink be $\beta(L_e + L_c/2)$. Then $(1-\beta)C_{bi}$ goes to C_L and βC_{bi} goes to C_2.

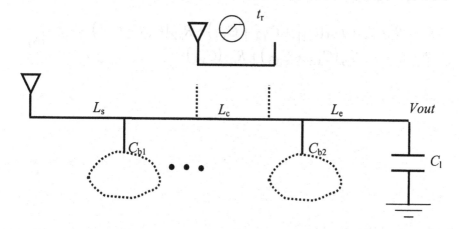

Figure 7-4. Extension of the 2-π model for general RC trees

7.4 MULTI SEGMENT NETS CROSSCOUPLING NOISE MODEL

As stated before, a net can have multiple horizontal and vertical segments. When we calculate the crosstalk noise in a net, we have to include the effect of all these segments and coupling to any of them. In this section, we show the extension to the noise model used to handle multiple segments, and in the next section we show how to handle multiple crosstalk for one segment.

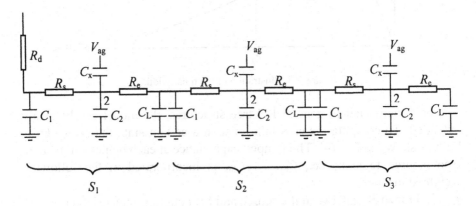

Figure 7-5. Multiple segment net model

For the example shown in Fig. 7-5, we have three segments connected to form a net. These segments can be either horizontal or vertical segments. We draw them as horizontal segments for clarity. The calculation for t_x, t_v for these three segments is shown below.

$$t_x = R_d C_{x1} + R_{s1}(C_{x1} + C_{x2} + C_{x3}) + R_{s2}(C_{x2} + C_{x3}) + \\ R_{s3}(C_{x3}) + R_{e1}(C_{x2} + C_{x3}) + R_{e2}(C_{x3}) \tag{6}$$

$$t_v = R_d \begin{pmatrix} C_{11} + C_{21} + C_{L1} + C_{x1} + C_{12} + C_{22} + \\ C_{L2} + C_{x2} + C_{13} + C_{23} + C_{L3} + C_{x3} \end{pmatrix} +$$

$$R_{s1} \begin{pmatrix} C_{21} + C_{x1} + C_{L1} + C_{12} + C_{22} + C_{x2} + C_{L2} + \\ C_{13} + C_{23} + C_{x3} + C_{L3} \end{pmatrix} +$$

$$R_{e1} \begin{pmatrix} C_{L1} + C_{12} + C_{22} + C_{x2} + C_{L2} + C_{13} + C_{23} + \\ C_{x3} + C_{L3} \end{pmatrix} +$$

$$R_{s2} (C_{22} + C_{x2} + C_{L2} + C_{13} + C_{23} + C_{x3} + C_{L3}) + \qquad (7)$$
$$R_{e2} (C_{L2} + C_{13} + C_{23} + C_{x3} + C_{L3}) +$$
$$R_{s3} (C_{23} + C_{x3} + C_{L3}) +$$
$$R_{e3} (C_{L3})$$

Where $C_{L1} = C_{e1}/2$, $C_{L2} = C_{e2}/2$, and $C_{L3} = C_{e3}/2 + C_l$
The general case where we have $h+v$ segments is shown below.

$$t_x = R_d \left(\sum_{i=1}^{h+v} C_{xi} \right) + \sum_{i=1}^{h+v} \left(R_{si} \sum_{j=i}^{h+v} C_{xj} \right) + \sum_{i=1}^{h+v-1} \left(R_{ei} \sum_{j=i+1}^{h+v} C_{xj} \right) \quad (8)$$

$$t_v = R_d \sum_{i=1}^{h+v} (C_{1i} + C_{2i} + C_{xi} + C_{Li}) +$$

$$\sum_{i=1}^{h+v} \left(R_{si} \sum_{j=i}^{h+v} (C_{2j} + C_{xj} + C_{Lj} + C_{1(j+1)}) \right) \qquad (9)$$

$$+ \sum_{i=1}^{h+v} \left(R_{ei} \sum_{j=i+1}^{h+v} (C_{2j} + C_{xj} + C_{L(j-1)} + C_{1j}) \right)$$

Where $C_{Li} = C_{ei}/2$ (i=1, ..., h+v-1) and $C_{L(h+v)} = C_{e(h+v)}/2 + C_l$

7.5 MULTI CROSSCOUPLING NOISE MODEL

We have extended the noise model to handle more than one aggressor at the same time. We consider aggressors that are directly adjacent to the victim. Aggressors that overlap with the victim but are not adjacent to it have no crosscoupling capacitive noise, as the wires between them act like shields. However, as discussed in previous chapters, these aggressors may have inductive coupling noise effects on the victim.

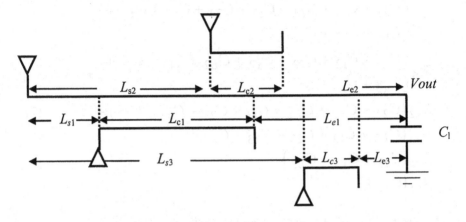

Figure 7-6. A victim with multiple crosscoupling aggressors

Fig. 7-6 shows a victim with multiple aggressors from both directions, above or under the victim (to the left or to the right for a vertical segment), and with different coupling lengths and locations. The way we can calculate R_s and R_e is to take the weighted sum of the crosscoupling capacitances to get the effective $(L_s+L_c/2)$ wire length. Having this effective length, all of the parameters like C_1, R_s, C_2, R_e, and C_L can be easily calculated. Eq. 10 show this effective $(L_s+L_c/2)$ calculation for one segment S_j which belongs to net N_i. S_j may have m different multiple crosscoupling as shown in Fig. 7-6.

$$Effective\left(L_s + L_c / 2\right) = \sum_{k=1}^{m} \left(\left(L_{sk} + L_{ck} / 2\right) * C_{xk}\right) / \sum_{k=1}^{m} C_{xk} \qquad (10)$$

The calculation of t_x is much more straight forward. We get C_x multiplied by its upstream resistance. We still can use the effective $(L_s+L_c/2)$ length multiplied by the total C_x.

7.6 WIRE SPACING

In this section, we just list two previously presented lemmas[50, 100]. The first one states that the peak noise monotonically increases as C_x increases. The second lemma states that the peak noise monotonically decreases as wire spacing increases, under the monotone capacitance model[50]. These two lemmas are of great importance to our work. Using these two lemmas, we can use any of the search methods like binary search or golden section search to find the minimum of the peak noise function. The noise function will be the v_{max} in (2), with the spacing as the varying parameter we optimize for.

7.7 POST LAYOUT RE-SPACING ALGORITHM

We aim to redistribute the spacing among wire segments in order to minimize the maximum noise that can appear on a path from the driver to any of the sinks in any net. As mentioned before, we use the same routing resources available; i.e., we do not increase the routing area, yet, we redistribute the spacing among wires to reduce crosstalk noise. The algorithm is an iterative one in which horizontal and vertical segments in any general layout routing can be respaced. It guarantees that the final layout is better or similar to the initial one. Preliminary results for this algorithm have been presented[101].

7.7.1 Algorithm

To alleviate the ordering problem in spacing distribution problems, we consider the paths in decreasing order of crosstalk noise. For each path, we replace (reposition) a segment, selected by either of the following two methods, to decrease the noise in that path without making any other paths more critical than the one under consideration. The two methods to select a segment to be replaced are: first, select the one which contributes the higher noise; second, consider the segments one by one starting from the driver side until reaching the sink side. The algorithm sorts the paths at the end of every iteration. The algorithm restarts a new iteration as long as we get some improvement compared to the previous iteration. The flow of the respacing algorithm is in Fig. 7-7.

1. **Get technology constants**
2. **Read layout**
3. **Build data structure for horizontal and vertical segments**
4. **Calculate the slack for each path = max crosstalk noise allowed in that path – current path crosstalk noise.**
5. **The main iteration is started.**
    5.1. Unlock all segments (all segments are allowed to move)
    5.2. For each path taken in increasing order of slack.
        1. **Go through each segment in the current path in order of decreasing crosstalk.**
        2. **If the segment is unlocked, allow the segment to move to increase path slack.**
        3. **Build the noise evaluation function by getting coupling information**
        4. **A segment is moved only if its move improves the slack in the path containing it and does not worse the slack in any other paths with slack less than that of the current path.**
        5. **Call the golden_section function to get the optimal location to move the segment to.**
        6. **Lock this segment.**
6. **If new slack = slack (for all paths), then quit, else continue.**
7. **Set slack equal to new slack.**
8. **Sort the paths' slacks again.**
9. **Go to 5.**

Figure 7-7. Pseudo code for post layout respacing algorithm

7.7.2 Complexity analysis

As it is clear from the algorithm pseudo code, the complexity of the respacing algorithm is the number of iterations, $\Im$ multiplied by the cost of each iteration. The complexity of each iteration includes paths' slacks sorting, crosstalk calculation for each segment, and golden section function evaluation for each segment to be replaced. The locking mechanism allow any segment to be moved (replaced) only once in each iteration. The complexity is $O(\Im \sum_{i=1}^{n} \sum_{j=1}^{h_i+v_i} p^{(i,j)})$, where $p^{(i,j)}$ is the cost to move segment j in path i to its optimal location.

The golden section search determines the optimal location for a segment in $O(t_x^{(i)} \lg(m/\varepsilon))$, where ε is the constant manufacturing element, m is max(height, width) of the routing area considered, and $t_x^{(i)}$ is the time required to calculate the new noise value in path p_i and is $O(h_i^2 + v_i^2)$ as we calculate *RC* for all the segments in a path that may be coupled to other segments. Putting all this together, $p^{(i,j)}$ is $O(h_i^2 + v_i^2 \lg(m))$.

7.8 EXPERIMENTAL RESULTS

We implemented the algorithm using C++ and tested it with some benchmark layouts on "Sun Ultra SPARC-III+" 900 MHz CPU workstation. Tabel 7-1 presents the specifications of the benchmarks used. We used the same benchmarks[98]. Gl4.opt and Deutsch are the two routings published before[102]. YK1 and YK4b are the published routings[103].

Table 7-1. Routing specifications for the benchmarks

Benchmark	#Nets	#Tracks	#Columns
GL4.opt	10	5	20
Deutsch.opt	72	19	174
YK1	21	12	37
YK4b	57	17	119

For the Gl4.opt example, we considered that the initial spacing between columns is 500μm and the spacing between tracks is 1.0 μm. Since the benchmarks available are girded, we assumed these initial spacing values. For the rest of the examples we assumed that the spacing between columns and the spacing between rows is α. We tried four different values for α, α=1.5 μm, 1.8 μm, 2.0 μm, and 0.2 μm.

Table 7-2 presents the initial and final noise for the layouts and the percentage of improvement (reduction in peak noise). As it can be seen in the table, we got 20% improvement for the GL4 example. The layout is shown in Fig. 7-8. The solid lines show the original location for the segments and the dotted lines show the new locations of the segments after applying the respacing algorithm for it. For this example, the horizontal segment in the very left top net could have been moved up. But, since we minimize the maximum noise and the noise in this net is not the maximum, there was no need to move that horizontal segment all the way up to the top.

Table 7-2. Noise reduction and improvement for the benchmarks

Benchmark	Initial noise Final noise (% improvement)			
GL4.opt	0.0255308			
	0.0206529 (19%)			
	α=0.2 μm	α=1.5 μm	α=1.8 μm	α=2.0 μm
Deutsch.opt	0.0900471	0.363538	0.426842	0.469044
	0.0885202(2%)	0.35447(3%)	0.416087(3%)	0.45724(3%)
YK1	0.0177688	0.0276151	0.0298873	0.0314022
	0.0177569(1%)	0.0259443(6%)	0.0270697(10%)	0.028429(10%)
YK4b	0.0954517	0.0972379	0.108963	0.11678
	0.0875645(9%)	0.0824468(18%)	0.0978632(12%)	0.0922456(26%)

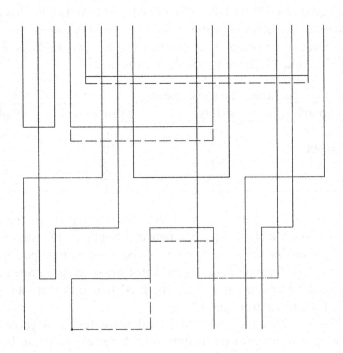

Figure 7-8. GL4.opt layout example

For YK1 and YK4b examples, we were able to get improvements ranging from 1% to 26%. The important thing about the improvements we get is that we do not consume any more resources. We just redistribute the spacing among adjacent wires. All it takes is more CPU time for the optimization in contrast to other techniques like sizing or shield insertion. For the Deutsch difficult example, we were able to get approximately 1% improvement even this example was routed to optimize for the crosstalk noise.

7.9 SUMMARY

We presented a re-spacing framework to reduce crosstalk noise. We explained the noise model used and how it is extended to handle simultaneous multiple crosscoupling. The algorithm and the model used can handle the effect of re-spacing among signals in one layer and the other connected segments in other layers. We presented a simple iterative algorithm that processes critical paths in decreasing order of crosstalk noise trying to minimize the noise in critical nets. We reported the complexity of our algorithm and results for some benchmarks.

Chapter 8

3D INTEGRATION

3D integration refers to the technologies that stack multiple levels of active devices in a vertical manner together with vertical interconnects between levels. 3D integration is pointed out by the International Technology Roadmap for Semiconductors (ITRS) as one of the most promising solution to sustain the performance improvement beyond 65 nm. It is one of the most promising solutions for achieving high-density device packaging and interconnects. One advantage of this method is that it can be used not only to minimize the distance that global interconnects must span, but also to provide integrated repeaters within the 3D layers to facilitate high-speed signal transmission.

In System on Chip, SOC, improved designs come at the cost of increased chip size and wiring. By layer stacking and having functional connectivity in the vertical direction, 3-D integration can provide similar functionality in a smaller footprint with reduced wiring lengths, thus reducing interconnect-driven signal delay and power consumption.

Many commercial products have demonstrated advantages of 3D integration technology.

8.1 EXISTING 3D INTEGRATION TECHNOLOGIES

Existing technologies can be classified into wafer-scale integration and packaging-based integration.

8.1.1 Wafer-scale integration

In the *wafer bonding approach*, multiple device wafers are bonded to each other with adhesives and level-to-level communication is provided by inter-level vias. Fig. 8-1 depicts the definition of a 3-D circuit, in which two device layers are both bonded and electrically interconnected using Cu-Cu pads (the bonding interface)[105]. Different schemes have been proposed with different bonding medium, wafer-to-wafer alignment accuracy requirement, and inter-level interconnection method.

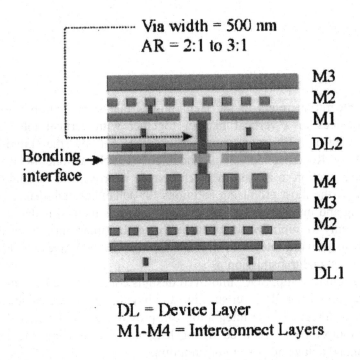

Figure 8-1. A typical 3D circuit

8.1.2 Packaging-based integration

One conventional technology is called *Stacked Package*, which stacks individually packaged chips (usually memory) on top of each other[106, 107]. Another technology is the *Stacked Chip-Scale Package* (S-CSP), which stacks bare dies and wire-bonds them to pads on the package substrate[108].

8.2 COMMERCIAL 3D DEVICES

Many commercial products have demonstrated advantages of 3D integration technology. These advantages are the small size, light weight, low cost, less noise issues, short interconnects, less power consumption, fast, and more integration into a single IC.

Matrix Semiconductor (Santa Clara, Calif.)[109] has developed what it says is the world's smallest 1 Gb silicon memory at 31 mm^2. Up to four cells are stacked on top of one another[110]. It is manufactured in a standard CMOS fab using existing semiconductor materials and production equipments. First, Matrix uses a standard 0.15 μm CMOS process for the logic that will drive the memory and interface functions. Above the CMOS, tungsten routing layers provide the interconnect wiring. Next, the polysilicon memory layers are created. Finally, an aluminum top metal is laid down for power and ground distribution. Unique to a 3-D integrated circuit, Hybrid Scaling is the combination of different process geometries within the layers of a 3-D circuit. This approach takes Matrix 3DM's leading memory densities even further than predicted by Moore's Law.

Irvine Sensors designed the computer stacks[111], confirming the feasibility and potential of the technology. A single stack is a complete computer with 52 chips on 48 layers, and includes 10 chip types. This includes a processor, interface chips, DRAM, and 32 layers of Flash memory for mass storage, eliminating the need for a hard drive. All of this is in an envelope that measures 0.8" x 0.5" x 0.5" high. Also included in the stack are precision resistors and capacitors, fabricated on silicon and processed in the stack in the same way as a chip. This design demonstrates that complex embedded systems requiring a high level of interconnectivity can be designed as Neo-Stacks.

Terrazon Semiconductor[112] has implemented an 8051 microcontroller IP core from CAST, Inc.[113], using its proprietary wafer stacking technology. The 3D IC processor Terrazon has created uses the vertical connections and stacking of their FaStack process to place 128 Kbytes of SRAM memory above the 8051-type processor and bind the two layers into a single device. Using a 160 nm CMOS process, the company claims that the resulting 3D IC runs up to ten times faster and requires only about 1/10th the power of a typical 8051. The core, CAST's R80515, is a reduced instruction set implementation of the standard 8051 Instruction Set Architecture. The FaStack process integrates multiple wafer layers and uses through-silicon vertical connections with the aim of creating 3D chips that are denser and faster than either traditional IC fabrication techniques or System-In-Package components. The company uses a special wafer thinning process that solves the thermal build-up problem of previous stacking technologies.

8.3 3D IC DESIGN TOOLS

The increased design complexity is a limitation of 3-D technology. 3D design and layout tools are the solution to cope with the complexity.

8.3.1 3D floorplanning

The existing works on 3D floorplanning can be divided into two approaches according to the 3D block representation. The first approach is to extend the existing 2D representation to a 3D representation, such as 3D TCG[115] and sequence triple[116], with a similar representation structure, a transitive graph and a sequence, respectively, added for the third dimension position representation. Since the number of device layers are limited, this approach is not suitable for 3D IC floorplanning as the additional complicated data Structure for the z-axis generates too much redundancy, which is not efficient in both time and space. The second approach for 3D floorplanning is to keep an array of 2D representations (2D Array), each representing all blocks located in one device layer, such as sequence pair[117]. The solution perturbations in these algorithms are limited by the 2D array-based representation. Very few effective z-axis moves can be enumerated, which limits the solution search space.

A thermal-driven 3D floorplanning algorithm that has a new 3D floorplan representation is proposed[118]. A 3D floorplan representation, combined bucket and-2D-array (CBA) is proposed with new solution perturbations added for 3D floorplanning. A compact resistive thermal model is integrated with the 3D floorplanning algorithm to achieve temperature optimization. Also, a thermal-driven 3D floorplanning algorithm based on a simplified closed-form thermal model is proposed for faster solution generation. Finally, a hybrid thermal model based on the two models mentioned above is proposed to get a good tradeoff between runtime and quality. The CBA representation is composed of two parts, a 2D floorplan representation is used to represent each layer, and a bucket structure to store the relationship between blocks at different device layers. Any 2D floorplan representation can be used. In order to encode the z-axis neighboring information, a bucket structure is posed on the circuit stack. In each bucket i, indexes of the blocks that intersect with the bucket are stored, the index set is referred to as $IB(i)$, no matter on which layer the block is located. In the meantime, each block j stores indexes to all buckets that overlap with the block, the index set is referred to as $IBT(j)$. Figure 8-2 illustrates an example with 7 blocks, a, b... g and two device layers, LI and L2, each represented by a separate TCG. A

2x2 bucket structure is imposed to the two layers with indexes listed in the buckets.

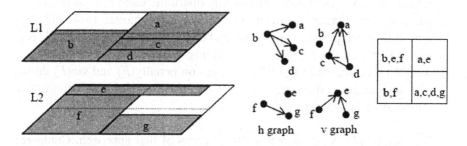

Figure 8-2. CBA of a Z-layer 7-block floorplan

A new interconnect-centric multi-layer floorplanner named Multi-layer Floorplanning System-on-Package, MF-SOP, which is based on a multiple objective stochastic Simulated Annealing method, is proposed[117]. The contribution of that work is first to formulate this new kind of floorplanning problem and then to develop an effective algorithm that handles various design constraints unique to SOP. The related experiments show that the area reduction of MF-SOP compared to its 2-D counterpart is on the order of $O(k)$ and wire length reduction is 48% average for k-layer SOP, while satisfying design constraints. Five types of moves for solution perturbation during Simulated Annealing were used: M1 (swap two modules in positive sequence), M2 (swap two modules from both positive and negative sequence), M3 (rotate), M4 is similar to M1, except that the two blocks are from positive sequences in different layers, and M5 selects a block from layer i and moves it to another layer j. The location in positive and negative sequence is randomly chosen.

A SOP floorplan F is *feasible* if (i) F is free of overlap among block location, (ii) F satisfies the layer and geometric constraints specified by the user. The goal is to minimize the following cost function based on an SOP floorplan solution F: $C(F) = c1 \cdot area(F) + c2 \cdot wire(F) + c3 \cdot via(F) + c4 \cdot penalty(F)$. The first term $area(F)$ is the final footprint area of SOP package. The minimization of this objective results in a minimal overall SOP package area. The second term $wire(F)$ is the half-perimeter bounding box (HPBB) based estimation of wirelength. The z-dimension has a direct impact on $via(F)$. If a net n spans from layer i to layer j, then $via(n) = |i - j|$. The sum of $via(n)$ for all nets is $via(F)$. The penalty term $penalty(F) = 0$ when there is no constraint violation in F. Their strategy for effective solution space search during Simulated Annealing is as follows:

1. construction of initial solution: first assign all blocks under layer constraints to the target layers and fix them during the annealing. For the remaining blocks, randomly and evenly distribute them into all layers.
2. solution perturbation: perform more inter-layer moves (M4 and M5) during high temperature annealing and more intra-layer moves (M1, M2, and M3) during low temperature annealing.
3. weighting constants in $C(F)$: focus more on penalty(F) and via(F) during high temperature annealing and more on area(F) and wire(F) during low temperature annealing.

The results obtained show the effectiveness of that approach. Compared to the single layer floorplanning, the final package area for 4-layer floorplanning is reduced by 73% on the average for the benchmarks used. The wirelength reduction for 4-layer floorplanning is 35% on the average compared to the single-layer case. The impact of geometric constraint on via results was not significant – only 4% average increase. The runtime has been increased by 9X with 4-layer floorplanning. The runtime slightly increased when MFSOP considers geometric constraints.

8.3.2 3D placement

The investigation and the adoption of 3D integration technology are done to increase the number of nearest neighbors for a module. In order to get the benefits meant by this technology, efficient placers tools are required. Typical 2D placer techniques include quadratic placement, simulated annealing, and min-cut portioning placement. With the advent of portioning algorithms that yield high quality solutions, placement by min-cut has become very attractive. In addition, the nature of the 3D integration technology can be seen as different partitions. Each partition is an active device level or plan.

3-D placement is solved by iteratively modifying a placement solution inside the unit cube[120]. The Gravity placement algorithm has four simple stages. The first stage is a random placement of nodes into the unit cube. This is followed by a force-based iteration that moves neighboring nodes closer together. After a number of forcebased iterations, node positions are rescaled in stage three to achieve an approximate uniform node distribution. After a number of repetitions of stages two and three, stage four determines the final placement through a recursive partitioning phase based on the nodes' computed coordinates.

A technique to reduce both local and global congestion in a 2.5D chip in order to increase the routability of the chip is provided[121]. They improve the temperature profile of the circuit using state-of-the-art thermal Alternate Direction Implicit, ADI, methodologies. The approach involves a two-stage refinement procedure: Initially, they use a multilevel min-cut based method to minimize the congestion and power dissipation within confined areas of the chip. This is followed by a simulated annealing-based technique that serves to minimize the amount of congestion created from global wires as well as improve the temperature distribution of the circuit. They showed that their congestion and thermal gradient minimization does not have any significant negative impact on the inter-layer wirelength or the number of intra-layer vias. They provided data to establish that these objectives are pairwise independent in the sense that minimizing the thermal gradient, for example, does not necessarily have a positive impact on total wirelength or wire congestion.

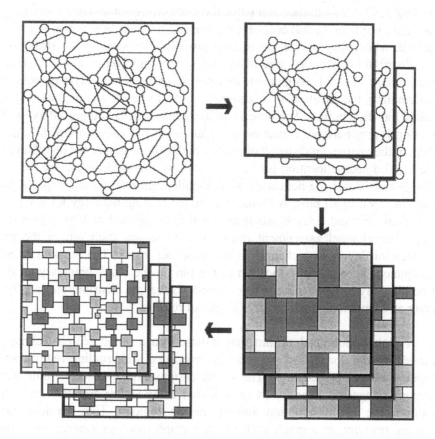

Figure 8-3. 3D physical design automation[121]

In summary, 3D placement tools should pay attention to constraints specific to 3D integration technology. The vertical interconnect, which is called Inter Chip Vias, will prevent the placement of active elements at these Inter Chip Vias positions. The placement tool may also account for buffer planning. Buffer insertion is a well-known technique to reduce interconnect delay and noise in deep submicron. Large numbers of buffers are needed and resources for these buffers should be planned ahead in early stages. In 3D technology, as stated by ITRS, integrated repeaters within the 3D layers are provided. Buffers can be placed in the active device level closer to the interconnect layer of the net segment being considered.

8.3.3 3D global routing

The net and pin distribution problem for global routing targeting three dimensional packaging layout via System-on-Package (SOP) has been studied[122]. They formulate and solve the multi-layer net and pin distribution for layer, wirelength, and crosstalk minimization. Their SOP router is multi-phased, where they divide the routing process into coarse pin distribution, net distribution, detailed pin distribution, topology generation, 2D layer assignment, channel assignment, and pin assignment step. The process of determining the location of entry/exit points for each routing interval is called pin distribution. The process of assigning nets to routing intervals (= routing layers between a pair of floorplan layers) is called net distribution step. In the coarse pin distribution step, which is done before net distribution, they find a coarse location for the pins and use this information for the net distribution. After the net distribution, detailed pin distribution step assigns finer location to all pins. A Steiner tree based routing topology for each net is constructed and a layer pair is assigned to it during topology generation step. The channel assignment problem is to assign each pin in the pin distribution layers to a channel in the floorplan layers. The purpose of pin assignment is to assign a location to the pin on the block boundary on the floorplan layer. They have implemented coarse pin distribution and 2-D layer assignment algorithms for SOP global routing[123].

A global routing algorithm that is based on a graph search technique guided by the congestion information associated with routing regions and topologies was presented[124]. The router assigns higher costs to route nets through congested areas (or those of higher delay and/or crosstalk costs) to balance the net distribution among routing regions. They modeled the routing resource as a graph such that the graph topology can represent the chip structure. Figure 8-4 illustrates the graph modeling. They first partition a chip into an array of rectangular subregions. These subregions are called global cells (GCs). A node in the graph represents a GC in the chip, and an edge denotes the boundary between two adjacent GCs. Each edge is assigned a capacity according to the physical area or the number of tracks of a GC. That work incorporated the features of having a new framework of performing congestion-driven *global* routing at the coarsening stage, followed by an intermediate stage of routing *layer/track assignment* for crosstalk optimization, and then *detailed* routing at the uncoarsening stage. By performing detailed routing after layer/track assignment, they can preserve more flexibility for allocating nets for crosstalk optimization. The minimum-radius minimum-cost spanning-tree (MRMCST) heuristic is adopted to construct routing trees for performance optimization[125]. An efficient and effective layer/track assignment scheme is incorporated for crosstalk and runtime optimization. Figure 8-4 shows their multilevel framework.

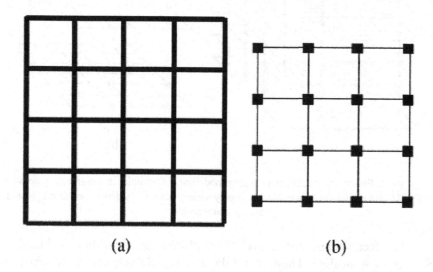

(a) (b)

Figure 8-4. Routing graph. (a) Partitioned layout. (b) Routing graph

8.3.4 3D power noise and thermal reduction

Efficient algorithms for 3D power supply noise, specifically the Simultaneous Switching Noise, SSN, and congestion analysis to guide the 3D module placement process is provided[126]. The goal in that work is to perform simultaneous SSN and congestion-aware physical design for 3D System-On-Package, SOP. In that approach, they first perform automatic placement of functional modules while minimizing the amount of decap required for each module. They then place decaps nearby the modules that need them while minimizing the overall footprint area. They developed a compact 3D SSN model that can be used to calculate the decap demand of a 3D solution efficiently. Their 3D wire congestion analysis consists of 3D pin redistribution and topology generation. Since 3D global routing update is time-consuming, they use trajectory-based method[127] to accurately estimate the congestion during Simulated Annealing process.

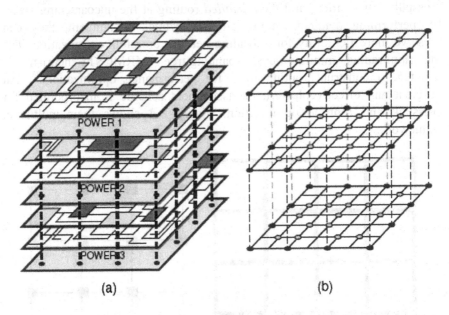

(a)   (b)

Figure 8-5. Illustration of 3D power supply modelling. (a) multilayer power supply network, (b) 3D-grid modelling, where black and gray nodes respectively denote power supply and consumption nodes[127]

The decap/congestion-aware SOP placement algorithm is based on Simulated Annealing. They extend the existing 2D Sequence Pair scheme to represent their 3D module placement solutions. The algorithm starts with a random placement solution. The placement is perturbed at each move and the cost is calculated. The cost involves area, wirelength, congestion, and decap. The moves are accepted based on improvement over previous cost or some probability dependent on the difference. The moves are made until freezing point is reached. The final placement is compacted by Linear Programming. They model the P/G network for 3D SOP as a 3D grid graph as shown in Figure 8-5. The edges in the grid-graph have inductive and resistive impedances. The mesh contains power-supply points and connection points, which supply and consume currents. The *dominant current source* for a block is defined as the voltage source supplying significantly more power to the block than any other neighboring sources. The *dominant path* for a block is the path from the dominant supply to the block causing the highest drop in voltage. It has been shown experimentally that the shortest path between the dominant current source (nearest Vdd pins) and the block offers highly accurate SSN estimation within reasonable runtime.

A very promising technique to mitigate thermal issues in integrated circuits is to lower the thermal resistance of the chip itself. This can be done by incorporating thermal vias into integrated circuits. This technique comes with the routing space overhead taken by the vias. Algorithms are needed to minimize the usage of the vias, place them in the most effective locations, and satisfy the design conditions. Because of the many dielectric layers, thermal problems are greater and thermal vias can have a larger impact in 3D ICs than 2D ICs. An efficient thermal via placement method was presented that attempts to overcome the thermal issues produced in the design of 3D ICs[128]. In order to make the placement of thermal vias more manageable, they reserve thermal via regions in certain areas of the chip for placing the thermal vias. A uniform density of thermal vias is used, and the thermal via placement algorithm determines the density in each of these regions. For a given placed 3D circuit, an iterative method was developed in which, during each iteration, the thermal conductivities of certain finite element analysis, FEA, elements are modified so that thermal problems are reduced or eliminated. These thermal conductivities reflect the density of thermal vias needed to be utilized within the element.

Chapter 9

EDA INDUSTRY TOOLS: STATE OF THE ART

The physical effects of deep submicron (DSM) interconnect are making it virtually impossible for front-end tools such as logic synthesis to predict what will happen during the back-end physical implementation process. This unpredictability can significantly affect a product's time-to-market. One of the most daunting tasks influencing delivery is achieving timing closure by bridging the gap between the front-end design task and backend implementation. There are two general approaches being proposed in the EDA industry to address the problem of timing closure – physical synthesis and physical optimization. In this chapter, we survey some tools in the EDA industry and outline its features and key benefits.

9.1 MENTOR GRAPHICS

9.1.1 TeraPlace

The Mentor Graphics TeraPlace™ tool suite is designed to deal with details at the physical level. The suite consists of the following solutions: TeraOptimize for physical optimization, TeraPlace™ for innovative timing-driven placement, and TeraCTS™ for clock tree synthesis. TeraPlace has both the algorithms and the architecture to achieve timing closure. To deal effectively with design complexity, TeraPlace is integrated with TeraOptimize and ·TeraCTS to provide a comprehensive solution for physical implementation. TeraCTS addresses the critical aspects of the clock structure by supporting gated and derived clocks and managing the clock skew. TeraOptimize deals with both timing and signal integrity issues by taking accurate physical data into account and using it to drive the optimization process. The full-featured timing analysis in TeraOptimize eliminates false and multi-cycle paths so both the optimization and timing-driven layout focus only on true timing paths. The delay numbers used by the timing analysis tool are based on extracted data from the virtual router. The extraction rules can be based on equations or table models from the final extraction tool. Besides being the basis of the parasitic information, the virtual router communicates with the placement algorithms to quickly identify and remove routing congestion or hot spots. The TeraSearch placement algorithm has the ability to balance the various constraints and design objectives that are part of DSM design, such as timing, area, routability, wire length and signal integrity. This functionality is based on de facto standards to integrate smoothly into the leading design flows. To quickly converge on timing every operation in TeraPlace supports incremental execution.

TeraPlace integrates an innovative placement algorithm with the virtual router to attack routing congestion, thus eliminating detour routes. The TeraOptimize architecture breaks the mold of the traditional placement environment by integrating virtual routing, extraction with delay calculation, timing analysis and optimization into a single solution. The TeraPlace family of products is also the first placement and optimization solution that supports incremental execution of each operation in this physical implementation flow. The benefit of incremental execution is a performance boost that results in shorter time-to-market. Each step necessary for achieving timing closure can work either incrementally or on the complete design.

TeraOptimize is integrated with TeraPlace to avoid design iterations during physical implementation. Buffer tree manipulation is a good example of the benefits of this integration. Synthesis algorithms can only estimate the required buffers and guess how the instances should be connected because they do not understand the final placement. TeraOptimize works with the placement algorithm to determine the location and number of buffers required to meet timing, and at the same time reduce congestion and wire length.

Mentor Graphics has agreements with both Cadence® and Synopsys® to use their formats, allowing TeraOptimize to fit seamlessly into the industry's dominant design flows, and effectively bridge the gap between the front-end design and the back-end implementation worlds. The main objective of TeraOptimize is timing closure that starts with speeding up the slowest paths. This is achieved by identifying true timing paths. Following timing-driven layout, the physical optimization techniques used are buffer tree re-synthesis, gate sizing and repeater insertion. To deal with the challenge of signal integrity, TeraOptimize focuses on three areas: crosstalk, noise, and power. The crosstalk effects on timing are handled by the delay calculator, which varies the delay based on congestion and the type of check being performed, either setup or hold. To alleviate noise problems, TeraOptimize uses gate sizing and buffer insertion. The most sensitive path for noise occurs when a signal with a fast input slope is next to a signal with a slow transition time. TeraOptimize introduces a specialized algorithm called PIP (performance-based incremental placement). The PIP technology combines optimization technology with the placement algorithm. Traditional optimization routines try to correct timing by adjusting the logic. PIP expands on this concept by considering both logic and incremental placement changes to achieve timing.

TeraCTS focuses on reducing clock skew, which is an important step in the right direction. Following TeraCTS, TeraOptimize uses clock propagation when optimizing for both setup and hold times. TeraCTS fully supports multiple clock domains, and has been designed for DSM clocking structures such as gated and derived clocks, and various muxing strategies. To deliver a specified skew number, TeraCTS combines buffer insertion, virtual routing and incremental placement during clock-tree synthesis. The result is an integrated system that provides accurate parasitic data to the TeraCTS analysis tool, which can handle this type of complexity.

9.1.2 HyperLynx

Mentor Graphics Corp. has added new functionality to the standard version of its HyperLynx pre- and postlayout pcb signal integrity simulation and analysis tool. The company has also released a turbocharged version of the tool for gigahertz pc-board designs. With multiple acquisitions in the pc-board area over the last two years, Mentor now offers three signal integrity (SI) solutions. Its high-end solution, Interconnectix, is offered as part of the BoardStation suite. Quad XTK, which came from Innoveda Inc., is a part of the PADS pc-board design flow and the Cadence Allegro design flow. Hyper-Lynx is a stand alone tool for all design flows.

To the latest release, HyperLynx 7.0, Mentor has added new functionality and ease-of-use features not available in the HyperLynx EXT version. Today, whether they want to or not, all pcb designers are having to deal with some degree of high-speed design. And over the last 14 to 18 months there has been a real move toward multigigabit, serialized, asynchronous design method and with it the needs for high-speed design are changing dramatically. What used to be a matter of delay, overshoot, undershoot, and crosstalk have now become fairly deep analog problems.

Furthermore, the company said it has improved differential signaling so that the tool can identify quick terminators for differential signaling. And the tool's oscilloscope feature now can show IC thresholds. HyperLynx now works with Mentor's BoardSim and ICX products and supports new Ibis models with V-I and Vt curve correction.

For SI analysis of designs above 500 MHz, the new HyperLynx GHz includes all the features of HyperLynx EXT, plus a number of advanced SI features that have been simplified for nonexpert users. The most notable addition is an inter-symbol-interference (ISI) diagram analysis feature, which provides views of time domains and switching data for gigahertz designs. Users can define multi-bit stimulus for a net, including toggle (clock), pseudorandom, 8B/10B encoding or custom. The tool also can simulate clock, signal propagation and receive/PLL jitter and overlay "eye mask" thresholds on software waveforms, instead of hooking a chip up to an oscilloscope to see if it works in the context of a given high-speed design layout. Related to the new eye diagram feature, Mentor has also included more-precise transmission-line and via models, which the company said are required for gigahertz designs. Many of these are created in Spice for use in Synopsys Inc.'s HSpice simulator. HyperLynx GHz also includes a function that allows users to control HSpice from the HyperLynx graphical user interface. The ISI function includes several built-in features that automate advanced HSpice subroutines under the hood. They deal with HSpice in the same easy-to-use HyperLynx environment, but they get that depth and power they need from the Spice simulator.

9.2 CADENCE

9.2.1 Virtuoso Layout Editor Turbo

Virtuoso® Layout Editor Turbo is the mid-range custom block authoring physical layout tool of the Virtuoso custom design platform. It supports the physical implementation of custom digital, mixed-signal, and analog designs at the device, cell, and block level. The Virtuoso custom design platform is a comprehensive system for fast, silicon-accurate design and is optimized to support "meet-in-the-middle" design methodologies such as advanced custom design. Virtuoso includes the industry's only specification-driven environment, multi-mode simulation with common models and equations, vastly accelerated layout, advanced silicon analysis for 0.13 microns and below, and a full-chip, mixed-signal integration environment. The Virtuoso platform is available on the Cadence® CDBA database and the industry standard Open Access database. With the Virtuoso platform, design teams can quickly design silicon that is right and on time at process geometries from one micron to 90 nanometers and beyond.

9.2.2 SPECCTRA® Expert

SPECCTRA® Expert Autorouter is the market's leading solution for automatic and interactive interconnect routing for printed circuit boards and complex IC packaging. Designed to handle high-density PC boards requiring complex high-speed design rules, SPECCTRA® Expert uses powerful shape-based algorithms to make the most efficient use of the routing area. The results are increased productivity and shortened design cycles. Some key features for SPECCTRA® Expert are the high-speed rules/constraint support for delay, crosstalk, impedance control, differential pair, and net scheduling – including virtual pins, real-time routing to net scheduling, delay, crosstalk, impedance, and differential pair rules, auto net shielding with auto via stitching, and true 45-degree or orthogonal routing.

Crosstalk is limited by geometric or parallelism rules. You can define the acceptable gap and length parameters and the autorouter automatically separates the parallel lines after a specified distance. Create a gap versus length table to more accurately model crosstalk. Crosstalk is reduced by evaluating the geometric or physical layout of conductors and the electrical properties of signals carried by the conductors. Noise associated with conductors both on the same layer and adjacent layers is considered. Cumulative noise crosstalk is controlled by coupled noise rules that incorporate the design's electrical properties. SPECCTRA® users can define the maximum cumulative crosstalk noise in millivolts allowed by each class or net. SPECCTRA® Expert autorouter dynamically computes the maximum cumulative crosstalk noise during routing. This is achieved by summing all parallel and tandem conductors.

SPECCTRA® also has the flexibility to handle the special geometry requirements of high-performance designs. For differential pair routing, users define the gap between the two conductors and the autorouter takes care of the rest. The autorouter intelligently handles routing around or through vias and automatically conforms to any defined minimum length criteria. Automatic net shielding is used to reduce noise on noise-sensitive nets. Separate design rules may be applied to different regions of the design. For example, you can specify tight clearance rules in the connector area of your design and less stringent rules in the rest of your design. With signal margins shrinking rapidly, the use of net shielding is becoming commonplace on most high-speed designs. In order for net shielding to perform correctly, a shield needs to be connected to the ground plane at regular intervals so that any stray noise can be effectively absorbed.

SPECCTRA® can not only auto-generate the shield around specified nets, it can also tie the shield route to the chosen ground plane by automatically inserting vias at specified distances.

The advanced rule set feature within SPECCTRA® provides the capability for electrical parameter control, reporting crosstalk, and conductor-length rule violations – requirements demanded by today's large designs. Electrical parameter controls include the flexibility to assign specific rules to each element in your design. Users define the rules required to meet the electrical class characteristics unique to each layer, via type, conductor width, or set of connections. Using this feature, a larger via can be used to support the increased current capacity required by power and ground connections. This conserves design space because only the selected signals use larger vias. Additionally, you can improve signal impedance matching by assigning different width and clearance rules to different layers. Outer layers generally have higher impedance and are assigned larger conductor widths. Inner layers generally have lower impedance and are assigned narrower widths to match the impedance of the outer layers. Impedance matching can also be improved by controlling routed nets on certain layer or layer pairs. Reports include coupled noise by using the geometric or physical layout of conductors and electrical properties of signals. Noise associated with signals on the same layer and on adjacent layers is considered. Possible timing problems are reported by showing minimum and maximum wire length violations. All this information is reported in a file and displayed graphically.

9.2.3 SPECCTRAQuest

Differential technology offers many advantages in the design of digital products that require fast communication between chip, package, and board. Reduced noise, faster circuit speed, and reduced power consumption are among them. At its simplest, differential signaling is a method of sending the same piece of information both non-inverted and inverted from a (differential) driver over two traces to a (differential) receiver. Differential signals allow for lower voltage swings resulting in faster circuits, reduced power consumption, and reduced electro-magnetic interference (EMI). These benefits are realized by design complexities and challenges that include: the assignment, grounding and powering of IC pins; exploring implementation parameters and routing traces accordingly in a tandem fashion; and matching layer assignments and locations of vias. Differential designs are developed and validated by simulation. The new Cadence® differential signaling technology is incorporated in the company's SPECCTRAQuest design and analysis solution. It allows users to simulate differential signals as a unit in pre- and post-layout simulations, improving both design quality and productivity. Key features and benefits of the differential signaling capability are:

With the new ability to see differential signals as an entity, sweep simulation allows users to explore both design parameters such as differential impedance and differential propagation delay, as well as physical implementation constraints such as trace width and gap, out-of-phase tolerance, and length or trace coupling tolerances.

Custom stimulus and custom measurement capabilities for differential signals allow users to make complex measurements such as common mode offset or source-synchronous timing measurements from die pads inside the IC package. A user can now understand the influence of the package on the performance of the differential signal.

Ability to extract differential signals from layout design for post-layout verifications and debugging purposes eliminates a number of time-consuming tasks.

By using the Cadence Advanced Package Designer database, users can simulate and verify differential signals from die-to-die through the PCB.

9.2.4 PacifIC

PacifIC is a static noise analyzer for custom digital Ics. PacifIC helps you avoid chip failures due to noise in advanced high-speed custom digital designs. It analyzes the combined impact of all major noise sources including crosstalk, charge sharing, leakage, IR drop, overshoot, undershoot, and propagated noise. PacifIC also supports analysis of signal-integrity problems unique to silicon-on-insulator (SOI) designs. PacifIC has some key features and benefits as it prevents potential chip failures due to noise, identifies noise sensitive circuitry to improve yield, performs advanced circuit and interconnect noise analysis, calculates the noise immunity of every node in the design, employs static analysis – no vectors required, supports hierarchical design using abstract noise models (ECHOs), integrates easily into existing design flows via standard data formats, and analyzes floating body and parasitic bipolar noise effects in SOI based designs. PacifIC analyzes all possible sources of noise acting together in the worst possible scenario. It can be used throughout the design process to manage the trade-off between noise margin, area, and performance. PacifIC can also be used for final verification sign-off. PacifIC contains an advanced proprietary Krylov subspace algorithm for efficiently analyzing large coupled networks integrated with non-linear transistor level transient simulation. In addition to interconnect noise, PacifIC calculates circuit noise due to charge-sharing and allows noise to propagate from previous logic stages. Timing information and logic analysis are used to remove pessimism. Noise sources are only combined when it is possible for them to act together. PacifIC automatically determines the noise immunity of every node in the design by means of a sensitivity-based noise stability metric. This metric helps localize failures near their source and guarantees adequate noise immunity for better design quality and yield. PacifIC is able to perform full-chip analysis through the use of high-level noise abstractions. PacifIC supports two models, a user defined noise (UDN) model and an ECHO™ model. UDNs are used for incomplete or non-digital blocks. ECHOs are created from transistor level noise analysis by PacifIC. PacifIC also uses a built-in noise aware macro model for analysis of long chip level interconnects. The combination of noise abstracts and interconnect macro-modeling enables PacifIC to analyze multimillion transistor designs. PacifIC supports standard SPICE, SPEF, and DSPF formats with BSIM3, BSIM4 and MOS9 device models as input. PacifIC outputs circuit snapshots in standard SPICE format. Optional inputs include signal timing windows, logic constraints and user-defined noise assertions. PacifIC is user configurable via Tcl.

9.2.5 CADENCE SE-PKS

Silicon Ensemble® PKS (SE-PKS) is the physical implementation cornerstone of Cadence® Synthesis/Place-and-Route and supports flows from RTL or gates to GDSII. In particular, SE-PKS optimization/place-and-route – the heart of the overall Cadence SP&R product line—is targeted for the design and implementation of leading-edge integrated circuits (ICs). In addition to its comprehensive conventional place-and-route features – floorplanning, placement, clock-tree synthesis, routing, extraction, and timing analysis – SE-PKS also includes Physically Knowledgeable Synthesis (PKS) concurrent optimization, power analysis, and signal integrity prevention, analysis, and correction capabilities. It also includes advanced features designed specifically to address manufacturing requirements for designing at the 130 nm process technology node. These and other features make SE-PKS the ultimate tool for leading-edge block-based design and implementation. Some of its benefits are the fastest path to tapeout with rapid timing and design closure, highest performance results from PKS concurrent optimization, unparalleled predictability due to consistent core technology engines through the flow, minimal iterations compared to wireload-based synthesis methodologies, best-in-class integrated signal integrity features enabling correct-by-construction design, handles all known 130 nm foundry requirements for next-generation design, multi-CPU routing for enhanced throughput capabilities, and lowest risk with over 15,000 tapeouts using its core technology.

SE-PKS can accept sub-optimal netlists from conventional wireload model-based synthesis tools and completely restructure them based on accurate physical information. It does so employing concurrent PKS optimization. PKS reads the inaccurate netlist and any physical constraints, and uses its set of powerful synthesis transforms to re-optimize the netlist using actual physical information. It optimizes the real critical paths, taking congestion information into account, and using true global routing to determine interconnect timing. As described earlier, it uses the same placement and routing engines in this stage as are used throughout the flow. The output of PKS is stage legally placed netlist that correlates to the final post-routing timing within 5%. The same is true starting from RTL instead of gates – but in the case of RTL, the results are even better because PKS takes advantage of many more degrees of freedom to derive its final results.

The analysis tools identify the SI problems but offers no automated way to fix them. Therefore, the designer is left to make manual changes to the design. This can be very difficult since after detailed routing, there is very little flexibility left for the designer to maneuver the placement and routing. These fixes can also potentially disturb other design constraints that were previously satisfied. This process, by its very nature, can be non-convergent and impact the time-to-market windows for products. SE-PKS promotes the signal integrity awareness to all the phases of the design cycle from synthesis down to detailed routing. The implementation and signal integrity analysis go hand in hand throughout the flow including various novel techniques to prevent the signal integrity issues in the first place. This solution is based on a broad scope of SI issues – timing, power, and area – which are all kept in mind as the designer navigates through the design flow. In essence, all technology engines in the design solution are SI aware and can predict and prevent most SI related issues before they become real. In those cases where prevention is not possible, SE-PKS determines automated ways of correcting the SI problems without disturbing other design parameters. To solve SI issues proactively, SE-PKS incorporates best-in-class crosstalk prevention, analysis, and correction capabilities as well as IR drop and electromigration analysis and correction. The seamless integration of CeltIC SI for crosstalk provides a SPICE-like accuracy for glitch and noise-on-delay calculations and contributes to a reduction of crosstalk repairs and false failures by 90 percent when compared to competitive solutions. This means that teams working with 130 nm process technologies can confidently deal with and/or eliminate any and all crosstalk issues during place-and-route instead of after the fact. By handling signal integrity issues automatically during physical implementation, design teams avoid post-place-and-route manual fixes and the resulting costly iterations.

9.2.6 CeltIC

CeltIC is an advanced crosstalk analyzer for digital CMOS ICs that calculates the impact of crosstalk on both functionality and delay. It analyzes and propagates glitch noise to verify noise immunity and ensure functional validity of the circuit. It also outputs noise induced delay changes in SDF format for feedback to static timing analysis. Additionally, it can repair crosstalk problems and generate ECOs for place-and-route. CeltIC handles multimillion gate SoC designs flat or hierarchically using ECHO models. CeltIC is seamlessly integrated with Cadence SoC Encounter and Cadence Silicon Ensemble® -PKS (SE-PKS), and is also available individually for use with third-party place-and-route, parasitic extraction, and static timing analysis tools. CeltIC supports standard library and interface formats. Some of its key benefits are:

- prevents silicon respins due to noise related functional failures
- accurately accounts for crosstalk effects on timing
- improves yield by fixing nets with low noise immunity
- reduces design iterations via early detection of signal integrity problems
- integrates seamlessly into the Cadence design flow

9.2.7 SIGNALSTORM

SignalStorm® offers a unified signal integrity solution for cell-based design that accounts for propagation delay, IR drop, and crosstalk effects. It addresses next-generation, hierarchical delay calculation to provide the industry's most advanced technology for the signal integrity timing issues that are your critical concern for aggressive 0.18 micron designs and mainstream 0.13 micron designs. Some of its benefits and features are:
- unified signal integrity solution for cell-based designs that accounts for both crosstalk and voltage (IR) drop in delay calculation
- the most accurate delay calculation for popular timing flows:
  - unique time-quantized calculation of C_{eff}
  - accurate multi-driven net and mesh handling
  - non-linear modeling of instance-specific IR drop
  - includes crosstalk-induced signal delays
- fully hierarchical delay calculation provides fastest, most efficient commercial delay calculation with capacity for large, complex designs
  - 4x to 20x faster delay calculation

- ▪ 80% less memory consumption
- • identifies glitch transitions induced by crosstalk
- • production-proven technology
  - ▪ includes advanced extraction technology
  - ▪ leverages instance-specific IR drop data from VoltageStorm®
  - ▪ used in dozens of high-performance, high-volume designs
- • optional, advanced driver model for aggressive, complex design styles
  - ▪ first commercial application of effective current source modeling (ECSM)
  - ▪ more accurately represents complex topologies
- • advanced, automatic library characterization
  - ▪ automatic function recognition
  - ▪ .lib and ECSM characterization
  - ▪ automatic SPICE job management
  - ▪ binary search-based timing-check characterization
- • compatible with industry-standard libraries and popular static timing analysis tools

SignalStorm incorporates several industry-first technologies to provide the most accurate delay calculation solution available today for the most popular timing analysis flows using the standard Liberty .lib format. Increased SignalStorm accuracy yields fewer false timing violations, conserving valuable engineering resources, speeding time-to-market and enabling more aggressive designs. First, SignalStorm employs a time-quantized calculation of the effective load capacitance of receiver gates. This unique technology computes the effective capacitance over several time steps during the signal ramp to more closely track the actual effective capacitance as it changes with signal voltage. SignalStorm also incorporates instance-based IR-drop data from VoltageStorm, to account for the effects of IR drop on path delays. This capability becomes critical at 0.13 micron and below, where power supplies are typically in the 1.2 volt range, and even small voltage drops can compromise signal timing and lead to chip failures. Unlike other delay calculators that treat IR drop as a simple voltage de-rating factor, SignalStorm models IR drop as a non-linear effect, avoiding the limited inaccuracy inherent in linear K-factors. SignalStorm represents multi-driven nets and meshes with great accuracy, supporting complex clock distributions. Finally, SignalStorm accounts for the effects of cross-coupling between simultaneously transitioning signals with corrections to both delay calculations. Using the "glitch report" produced by SignalStorm, you can also analyze unintended transitions due to cross-couplings. SignalStorm delay calculation is not only the most accurate available, it is also the fastest. An efficient, hierarchical database structure enables delay calculation that is

4 to 20 times faster while using only one-fifth the memory of other tools. A hierarchical database ensures any changes in child cells are automatically reflected in their parents. This unique, hierarchical data model enables SignalStorm to easily handle today's largest, most complex SoC designs— without the need for you to build timing model extracts.

SignalStorm integrates easily into standard timing verification flows. It can read library and design information from standard LEF/DEF and .lib formats using leading 3-D accurate extraction technology, which is embedded into SignalStorm, thus providing the best-possible parasitics for delay calculation. SignalStorm transfers the extracted RC data in binary format into the fully hierarchical database of the delay calculator, significantly reducing the amount of time needed for processing the RC files, and resulting in disk-space savings of more than 80%. Alternatively, SignalStorm can read standard DSPF and SPEF files produced by other extraction processes. SignalStorm accepts instance-based IR drop data from VoltageStorm to account for the impact of IR drop on path delays. SignalStorm outputs SDF files for use by industry-standard static timing analysis tools, and also produces text-based reports highlighting signal integrity risks.

9.2.8 NANOROUTE ULTRA

NanoRoute™ is a core nanometer technology included in the Encounter™ digital IC design platform – it offers a full placed gate-to-GDSII solution that delivers scalable performance and capacity for routing and optimizing large SoC, ASIC, and application-specific standard product (ASSP) designs. It is the industry's first and only unified routing and physical optimization solution that provides one-step rapid timing closure and signal integrity prevention, analysis, and correction *during* detailed routing. Some of its key benefits are:

- Intuitive, easy-to-use physical implementation system based on First Encounter® graphical user interface
- Complete physical implementation system from placed gates-to-GDSII
- All-purpose solution for block-level and chip-level routing
- Routing capacity of over 10-million gates flat and 25+-million gates hierarchical designs — the highest capacity router for both flat and hierarchical methodologies 10X and more speed-up and capacity increase over existing routers
- Cuts routing from days to hours and minutes
- Multi-CPU routing for even more performance boost
- Rapid timing and signal integrity closure

- Single-step routing and physical optimization for correct-by-construction routing
- Industry-standard inputs and outputs
- Adaptable to any design flow environment

9.2.9 DRACULA

Dracula® verification products are an established IC industry standard. You can trust Dracula to provide comprehensive and accurate sign-off verification results for all designs. This technology provides you with a complete set of verification tools suitable for small cells up to very large ICs. Dracula verification tools can be used no matter what your design methodology is – bottom-up, custom, standard-cell, structured gate array, or block-oriented. Dracula has produced industry-wide trusted results for over a decade. Some of its key benefits are:

- provides distributed processing support for reducing runtimes
- handles preverified or imported intellectual property (IP) blocks with its black box capability
- provides cross-probing and highlighting capabilities through its user-friendly GUI
- produces standard outputs for post layout simulation and signal integrity analysis

9.3 SYNOPSYS

9.3.1 Galaxy

Synopsys Inc. has demonstrated the most recent version of its register-transfer-level-to-GDSII integrated tool bundle, Galaxy, with new signal-integrity design and analysis capabilities woven into it. It has been introduced as an "RTL-to-GDSII design platform," Galaxy is a collection of point tools from Synopsys and its 2002 acquisition, Avanti, and includes Design Compiler, JupiterXT, Floorplan Compiler, Physical Compiler, Astro, StarRCXT, Hercules and Proteus. They are tied together with the Milkyway database and PrimeTime SI static-timing analysis tool. All tools in the environment use common libraries and constraint files, and designers have optimized the links among the various point tools to work cohesively.

In Galaxy SI, Synopsys has added signal integrity to the flow. They have had signal-integrity static crosstalk analysis, combining timing and static crosstalk into one tool with PrimeTime SI, for over a year. Now they are extending signal-integrity support to the full flow – the implementation and sign-off environments. Galaxy SI eliminates the effects of crosstalk noise, crosstalk delay, IR drop and IR drop's effect on timing throughout the flow. To provide that support, Synopsys added signal-integrity technologies to the Astro place and route and the Physical Compiler physical synthesis tools, added noise analysis to PrimeTime SI and resolved PrimeTime SI run-time problems. Designers had criticized PrimeTime SI v. 2003.03's run-times in a recent Synopsys Users Group (SNUG) survey on www.deepchip.com. The last version of PrimeTime SI took care of crosstalk's effect on delay, and now it can do glitch analysis. The company engineered a more consistent form of SI analysis by merging the Avanti tools with Synopsys tools.

Run-times have also improved with the most recent release. With PrimeTime SI, and with other tools for that matter, they focus on accuracy first and run-time performance second. Synopsys has also added a "placement-based" prevention feature to Physical Compiler Expert that prevents the placement technology from introducing crosstalk problems.

9.3.2 Astro-Xtalk

Astro-Xtalk, the signal-integrity option to Astro, offers the most comprehensive solution to signal-integrity issues in ultra-deep-submicron (UDSM) designs. Using the same timing engine, extraction engine, and optimization algorithm as used for timing and layout, Astro-Xtalk provides early visibility into signal integrity. Designers can take steps early on to prevent or correct potential signal-integrity problems during the physical-design process, not after. Astro-Xtalk's concurrent crosstalk analysis not only provides early indication of signal integrity aspects of the design, but also guides the physical implementation so that any prevention/correction action taken during the implementation does not generate another set of crosstalk problems elsewhere. The end result is a correct-by-construction physical design with signal integrity and timing closure achieved simultaneously. Astro-Xtalk is part of Synopsys' Milkyway™ -based, SinglePass-SoC™ solution. Some of its benefits are:

- enables designers to achieve signal integrity and timing closure concurrently, saving design time and ensuring maximum productivity.
- ensures signal integrity for UDSM designs with accurate crosstalk analysis and comprehensive crosstalk prevention and correction.
- speeds up time-to-market by eliminating signal integrity related problems, a primary cause of post-layout design iterations.
- saves engineering cost because of easy integration into existing design flows.

9.4 ACCELERANT NETWORKS INC.

9.4.1 AN5000

Accelerant Networks is a developer of high-integration ICs that allow rapid development of intelligent, high-speed backplane connection systems. To avoid signal integrity problems, the Accelerant design team needed a powerful verification methodology able to handle detailed simulations of a largely analog/mixed-signal design. Drawing on a combination of commercial and proprietary tools, the Accelerant team developed an effective mixed-signal verification flow, using full-chip, transistor-level, post-layout analysis. Accelerant Networks' signaling technology allows transceivers to operate in a backplane environment over copper on standard FR4 materials. At the heart of this new approach, multilevel analog signaling methods permit transmission of two data bits per clock — boosting data throughput while lowering the observed transmission rate. Deployed in new full-custom designs such as the Accelerant Networks AN5000, this type of signaling technology enables 5 Gbits/s bandwidth per differential pair using standard connectors and dielectric material for distances up to 48 inches.

With the AN5000 design effort, it is possible to move the entire design as a Spice netlist into HSIM (hierarchical circuit simulator) and perform full-chip simulations at full transistor-level accuracy. By simulating the large circuits at the transistor level, you can tune individual subcircuits with confidence not only in the accuracy of the target subcircuits but also in the accuracy of the remaining blocks. The final check in the Accelerant verification flow involves a comparison between pre-layout and post-layout results. By quantizing the post-layout analog waveforms to digital, they are able to compare those results with the Verilog digital output using timing window functions. This final step provides a critical check on timing of the interfaces. By their very nature, the post-layout HSIM simulations exhibit the effects of on-die signal integrity issues on dynamic timing. This is absolutely critical when routing large numbers of unrelated RF speed signals around the die.

9.5 SILICON METRICS

9.5.1 SiliconSmart SI

Silicon Metrics has added to its library development suite a new tool that performs signal integrity noise characterization and produces models for the latest generation of third-party signal integrity analysis tools. The new tool, SiliconSmart SI, addresses the growing need for tools that can accurately model serious signal integrity glitches, which are becoming more common in designs implemented with process geometries of 130nm and below. The new tool, based on Silicon Metrics' characterization technology and targeted toward library developers and CAD teams, allows users to generate accurate SI models, which are in turn analyzed and interpreted by popular signal integrity tools from companies such as Cadence Design Systems Inc. and Synopsys Inc. The tool generates models in various formats including Liberty for Synopsys' PrimeTime SI and ECSM for Cadence SignalStorm.

The company claims the latest signal integrity analysis tools can often produce hoards of data, much of it innaccurate, which makes it easy for design engineers to overlook show-stopping glitches. Thus the more accurate models produced by SilconSmart SI allows users of third-party SI analysis tool to home in on those glitches.

9.6 MAGMA

9.6.1 Diamond SI

RTL-to-GDSII design tool vendor Magma Design Automation Inc. expanded into the postlayout verification market, when it announced its Diamond SI signal integrity analysis tool. To date, Magma has offered an all-in-one physical-implementation tool suite – Blast Chip – that lets engineers do everything from register-transfer-level design planning through place and route. Now the company is rolling out a tool that does the next step in the IC development flow: signal integrity analysis. In keeping with Magma's all-in-one philosophy, Diamond SI analyzes timing; crosstalk noise and delay/glitch; voltage (or IR) drop; and electromigration – on both power and signal.

Diamond SI applies Spice-accurate transistor-level analysis to determine which nets have actual problems that must be corrected. The tool automatically generates the Spice decks for those critical nets and the necessary switching vectors for dynamic analysis. Its interface is compatible with all the big Spice and Spicelike simulators. The tool also outputs a report of problem areas, SPEF parasitics and SDF timing. Diamond SI uses resistance and capacitance (RC) extractor technology but does not currently find problems with inductance. It does analysis and directs users to problem areas but does not fix them or offer suggestions for rectifying problems.

References

1. http://public.itrs.net/Files/2001ITRS/Interconnect.pdf
2. H. B. Bakoglu. Circuits, Interconnections, and Packaging for VLSI, Addison-Wesley, 1990.
3. K. L. Shepard and V. Narayanan, "Conquering Noise in Deep Submicron Digital Design," IEEE Design Test Comput., pp. 51–62, Jan./Mar. 1998.
4. http://www.ece.iit.edu/~awang/seminar/00f/kang.html
5. Semiconductor Industry Association, International Technology Roadmap for Semiconductors, 1999, http://public.itrs.net/files/1999_SIA_Roadmap/Int.pdf
6. H. H. Su, K. Gala, and S. Sapatnekar, "Fast analysis and optimization of power/ground network," In Proc. Int. Conf. on Computer Aided Design, pp 477–480, 2000.
7. Shiyou Zhao, Kaushik Roy, and Cheng-Kok Koh, "Decoupling capacitance allocation for power supply noise suppression," In Proc. 2001 International Symposium on Physical Design, pp 66–71, 2001.
8. Lam, W.-C.D.; Cheng-Kok Koh; Tsao, and C.-W.A., "Power supply noise suppression via clock skew scheduling," International Symposium on Quality Electronic Design, 2002, pp 355–360.
9. http://developer.intel.com/technology/itj/q12001/articles/art_5.htm
10. B. K. Liew, N. W. Cheung, and C. Hu, "Projecting Interconnect Electromigration Lifetime for Arbitrary Current Waveforms," IEEE Trans. Electron Devices, vol. 37, no. 5, May 1990.
11. Banerjee, A. Amerasekera, N. Cheung, and C. Hu, "High-Current Failure Model for VLSI Interconnects under Short-Pulse Stress Conditions," IEEE Electron Device Lett., vol. 18, no. 9, September 1997.
12. Banerjee, A. Amerasekera, and C. Hu, "Characterization of VLSI Circuit Interconnect Heating and Failure under ESD Conditions," Int. Reliability Physics Symp., Dallas, TX, April-May 1996.
13. P. K. Chatterjee, W. R. Hunter, A. Amerasekera, S. Aur, C. Duvvury, P. E. Nicollian, L. M. Yang, and P. Yang, "Trends for Deep Submicron VLSI and Their Implications for Reliability," Int. Reliability Physics Symp., Las Vegas, NV, April 1995.
14. W. R. Hunter, "Self-Consistent Solutions for Allowed Interconnect Current Density--Part I: Implications for Technology Evolution," IEEE Trans. Electron Devices, vol. 44, no. 2, February 1997.

15. J. Cong, "An Interconnect-centric Design Flow for Nanometer Technologies," Int. Symp. VLSI Technology, Systems, and Applications, pp. 54–57, June 1999.

16. M. R. Stan and W. P, "Burleson. Bus-invert Coding for Low-power I/O," IEEE Trans. on VLSI Systems, pp. 49–58, Mar. 1995.

17. L. Benini, A. Macii, E. Macii, M. Poncino, and R. Scarsi, "Synthesis of Low-overhead Interfaces for Power-efficient Communication Over Wide Buses," ACM/IEEE Design Automation Conf., pp. 128–133, 1999.

18. P. R. Panda and N. D. Dutt, "Reducing Address Bus Transitions for Low Power Memory Mapping," European Design and Test Conf.," pp. 63–37, Mar. 1996.

19. H. Mehta, R. M. Owens, and M. J. Irwin, "Some Issues in Gray Code Addressing," the Great Lakes Symp. VLSI, pp. 178–180, Mar. 1996.

20. C. L. Su, C. Y. Tsui, and A. M. Despain, "Saving Power in the Control Path of Embedded Processors," IEEE Design and Test of Computers, vol. 11, no. 4, pp. 24–30, 1994.

21. L. Benini, G. De Micheli, E. Macii, D. Sciuto and C. Silvano, "Asymptotic Zero-Transition Activity Encoding for Address Busses in Low-power Microprocessor-based Systems," the Great Lakes Symp. VLSI, pp. 77–82, 1997.

22. E. Musoll, T. Lang, and J. Cortadella, "Working-zone Encoding for Reducing the Energy in Microprocessor Address Buses," IEEE Trans. on VLSI Systems, vol. 6, no. 4, pp. 568–572, Dec. 1998.

23. L. Benini, G. De Micheli, E. Macii, D. Sciuto, and C. Silvano, "Address Bus Encoding Techniques for System-level Power Optimization," Design, Automation and Test in Europe, Feb. 1998, pp. 861–866.

24. K.-W. Kim, K.-H. Baek, N. R. Shanbhag, C. L. Liu, and S. Kang, "Coupling-driven Signal Encoding Scheme for Low-power Interface-design," 2000 Intl Conf. on Computer-Aided Design, 2000.

25. S. Ramprasad, N. R. Shanbhag, and I. N. Hajj, "Information-theoretic Bounds on Average Signal Transition Activity," IEEE Trans. on VLSI, Sept. 1999.

26. R. Hegde and N. R. Shanbhag, "Energy-efficiency in Presence of Deep Submicron Noise," IEEE/ACM Int. Conf. Computer Aided Design, pp. 228–234, 1998.

27. Y. Zhang, W. Ye, and M. J. Irwin, "An Alternative Architecture for On-chip Global Interconnect: Segmented Bus Power Modeling," Asilomar Conf. on Signals, Systems, and Computers, pp. 1062–1065, 1998.

28. F. Catthoor, F. Franssen, S. Wuytack, L. Nachtergaele, and H. D. Man, "Global Communication and Memory Optimizing Transformations for Low Power Signal Processing Systems," VLSI Signal Processing VII, pp. 178–187, 1994.

29. W. Fornaciari, D. Sciuto, and C. Silvano, "Power Estimation for Architectural Exploration of HW/SW Communication on System-level Buses," Int. Workshop on Hardware/Software Codesign, pp. 152–156, 1999.

30. M. D. Powell, S. H. Yang, B. Falsafi, K. Roy, and T. N. Vijayku-mar, "Gated-Vdd: A Circuit Technique to Reduce Leakage in Cache Memories," Intl. Symposium on Low Power Electronics and Design, July 2000.

31. James T. Kao, and Anantha P. Chandrakasan, "Dual-Threshold Voltage Techniques for Low-Power Digital Circuits," IEEE Journ. Solid-State Circuits, vol. 35, July 2000, 1009-1018.

32. Assaderaghi Fariborz, Sinitsky Dennis, Parko Stephen, Bokor Jaffrey, Ko Ping, Hu, Chenming, "Dynamic Threshold-voltage MOSFET (DTMOS) for Ultra-low Voltage VLSI," IEEE Transactions on Electron Devices vol. 44, no. 3, March 1997.

33. Masayuki Miyazaki, Hiroyuki Mizuno, and Koichiro Ishibashi, "A Delay Distribution Squeezing Scheme with Speed-Adaptive Threshold-voltage CMOS (SA-Vt CMOS) for

Low voltage LSIs," 1998 International Symposium on Low Power Electronics and Design (ISLPE)

34. G. P. D'Souza, "Dynamic Logic Circuit with Reduced Charge Leakage," U.S. Patent 5483181, 1996.

35. J. J. Covino, "Dynamic CMOS Circuits with noise immunity," U.S. Patent 5650733, 1997.

36. L. Wang and N. R. Shanbhag, "Noise-tolerant dynamic circuit design," IEEE Intl. Symp. on Circuits and Systems, pp. 549-552, Orlando, FL, May/June 1999.

37. G. Balamurugan and N. R. Shanbhag, "Energy-efficient dynamic circuit design in the presence of crosstalk noise," ISLPED, San Diego, 1999.

38. G. Balamurugan and N. R. Shanbhag, "A noise-tolerant dynamic circuit design technique," 2000 Custom Integrated Circuits Conference, May 21-24, 2000, Orlando, FL.

39. Pant. M.D., Pant. P., Wills. D.S., Tiwari. V., "An architectural solution for the inductive noise problem due to clock-gating," International Symposium on Low Power Electronics and Design, pages 255–257,1999.

40. M. Graziano, G. Masera, G. Piccinini, M. Zamboni, "Automated Power Supply Noise Reduction Via Optimized Distributed Capacitors Insertion," Southwest Symposium on Mixed-Signal Design 2001, pages 162 – 167, 2001.

41. Kenneth L. Shepard and Vinod Narayanan "Conquering Noise in Deep-Submicron Digital ICs," IEEE Design & Test of Computers, vol. 15, issue 1, pp. 51 –62, Jan.-March 1998.

42. http://www.synopsys.com/products/tlr/flexroute_wp.pdf

43. Semiconductor Industry Association, International Technology Roadmap for Semiconductors, 2001, http://public.itrs.net/Files/2001ITRS/Interconnect.pdf

44. T. Sakurai, "Closed-form Expression for Interconnect Delay, Coupling, and Crosstalk in VLSI's," IEEE Trans. Electron devices, vol. 40, no. 1, pp. 118-124, Jan. 1993.

45. B. Kahng and S. Muddu, " An Analytical Delay Model for RLC Interconnects," IEEE Trans. On Computer-Aided-Design Integrated Circuits and Systems, vol. 16, no.12, pp. 1507-1514, December 1997.

46. Woojin Jin et al., "Experimental Characterization and Modeling of Transmission Line Effects for High-Speed VLSI Circuits Interconnects," IEICE Trans. Electron, vol. E83-c, no. 5, pp. 728-735, May 2000.

47. Mohamed Elgamel, Tarek Darwish, Magdy Bayoumi, "Noise Tolerant Low Power Dynamic TSPCL D Flip-Flops," IEEE annual symposium on VLSI, ISVLSI 2002, pp. 89-94, 2002.

48. Vital and M. Marek-Sadowska, "Crosstalk reduction for VLSI," IEEE Trans. Computer-Aided Design, vol. 16, March 1997.

49. L. T. Pillage and R. A. Rohrer, "Asymptotic Waveform Evaluation for Timing Analysis," IEEE Transactions on Computer-Aided Design of ICs and Systems, vol. 9 pp. 352-366, April 1990.

50. J. Cong, D. Z. Pan, and P. V. Srinivas, "Improved Crosstalk Modeling for Noise Constrained Interconnect Optimization," Proc. ACM/IEEE International Workshop on Timing Issues in the Specification and Synthesis of Digital Systems, pp. 14-20, Dec. 2000.

51. B. Kahng, S. Muddu, and D. Vidhani, "Noise and Delay Uncertainty Studies for Coupled RC Interconnects," Proc. IEEE International ASIC/SOC Conference, pp.3-8, Sep. 1999.

52. Ashok Vittal and Marek-Sadowska, "Crosstalk in VLSI Interconnections," IEEE Trans. Computer-Aided Design, vol. 18, no. 12, December 1999.

53. Jeffrey A. Davis and James D. Meindl, "Compact Distributed RLC Interconnect Models," IEEE Transactions on Electron Devices, vol. 47, no. 11, November 2000.

54. M. Kuhlmann and S. S. Sapatnekar, "Exact and Efficient Crosstalk Estimation," IEEE Transactions on Computer-AidedDesign of Integrated Circuits and Systems, vol. 20, no. 7, pp. 858-866, July 2001.

55. Andrew Kahng, Sudhakar Muddu, and Devendra Vidhani, "Noise and Delay Uncertainty Studies for Coupled RC Interconnects," proc. Twelfth Annual IEEE International ASIC/SOC Conference, pp. 3-8, 1999.

56. Andrew Kahng, Sudhakar Muddu, and Devendra Vidhani, "Noise Model for Multiple Segmented RC Interconnects," 2001 International Symposium on Quality Electronic Design, pp. 145-150, 2001.

57. Murat R. Becer, David Blaauw_, Vladimir Zolotov, Rajendran Panda and Ibrahim N. Hajj, "Analysis of Noise Avoidance Techniques in DSM Interconnects using a Complete Crosstalk Noise Model," proc. The 2002 Design Automation and Test in Europe Conference and Exhibition (DATE'02), pp. 456-463, 2002

58. Sachin S. Sapatnekar, "A Timing Model Incorporating the Effect of Crosstalk on Delay and its Application to Optimal Channel Routing," IEEE Trans. Computer-Aided Design, vol. 19, no. 5, May 2000.

59. Supamas Sirichotiyakul, David Blaauw, Chanhee Oh, Rafi Levy, Vladimir Zolotov, Jingyan Zuo "Driver Modeling and Alignment for Worst-Case Delay Noise," DAC 2001, June 18-22, 2001, Las Vegas, Nevada, USA.

60. Yehea I. Ismail and Eby G. Friedman, On-Chip Inductance in High Speed Integrated Circuits, Kluwer Academic Publishers, 2001.

61. Hai Zhou, D. F. Wong, I-Min Liu, and Adnan Aziz, "Simultaneous Routing and Buffer Insertion with Restrictions on Buffer Locations," IEEE Trans. Computer-Aided Design, vol. 19, no. 7, July 2000.

62. J. Cong and K.S. Leung, "Optimal Wire Sizing Under the Distributed Elmore Delay Model," Proc. Int. Conference on Computer Aided Design, pp. 634-339, 1993.

63. J. Cong, K.S. Leung, and D. Zhou, "Performance-Driven Interconnect Design based on Distributed RC Model," Proc. Design Automation Conference, pp. 606-611, 1993.

64. J. Cong, L. He and C.-K. Koh, "Global Interconnect Sizing and Spacing with Consideration of Coupling Capacitance," Proc. ACM/IEEE Int'l Conf. on Computer-Aided Design, November 1997.

65. J. Lillis, C. K. Cheng, and T. T. Y. Lin, "Simultaneous Routing and Buffer Insertion for High Performance Interconnect," Proc. the Sixth Great Lakes Symp. on VLSI, 1996.

66. Mohamed A. Elgamel, K.S. Tharmalingam, and Magdy A. Bayoumi, "Crosstalk Noise Analysis in Ultra Deep Submicrometer Technologies," IEEE Computer Society Annual Symposium on VLSI (ISVLSI'2003), February 20-21, 2003, Tampa, Florida, USA.

67. K. Chaudhary,A. Onozawa, and E. S.Kuh, "A Spacing Algorithm for Performance Enhancement and Cross-talk Reduction," in Dig. Tech. Papers IEEE/ACM Int. Conf Computer-Aided Design, pp. 697–702, Nov. 1993.

68. Hsiao-Ping Tseng, Louis Scheffer, and Carl Sechen, "Timing and Crosstalk Driven Area Routing," Design Automation Conference, pp. 378-381, 1998.

69. Mohamed Elgamel, K.S. Tharmalingam, and Magdy A. Bayoumi, "Noise-Constrained Interconnect Optimization for Nanometer Technologies," 2003 IEEE International Symposium on Circuits and Systems (ISCAS'2003), May 25-28, 2003, Bangkok, Thailand.

70. Lei He and Kevin M. Lepak, "Simultaneous Shield Insertion and Net Ordering for Capacitive and Inductive Coupling Minimization," proceedings of the International Symposium on Physical Design, May, 2000, San Diego, CA USA.

71. Kevin M. Lepak, Irwan Luwandi, Lei He, "Simultaneous Shield Insertion and Net Ordering under Explicit RLC Noise Constraint," proc. Design Automation Conference, pp. 199 –202, 2001.

72. J. S. Yim and C. M. Kyung, "Reducing Cross-Coupling among Interconnect Wires in Deep-Submicron Datapath Design," 36th Design Automation Conference (DAC), pp. 485-490, June 1999, New Orleans, USA.

73. C. -C. Chang, J. Cong, D. Zhigang, and X. Yuan, "Interconnect-Driven Floorplanning with Fast Global Wiring Planning and Optimization," Proc. SRC Techcon Conference, September 21-23, 2000, Phoenix.

74. Dennis Sylvester and Kurt Keutzer, "A Global Wiring Paradigm for Deep Submicron Design," IEEE Trans. Computer-Aided Design, vol. 19, no. 2, February 2000.

75. D. Sylvester and K. Keutzer, "Getting to the Bottom of Deep Submicron," in proc. ICCAD, 1998, pp. 203-211.

76. Jason Cong et al, "DUNE-A Multilayer Gridless Routing System," IEEE Trans. Computer-Aided Design, vol. 20, no. 5, May 2001.

77. Rony Kay and Rob A. Rutenbar, "Wire Packing-A strong Formulation of Crosstalk-Aware Chip-Level Track/Layer Assignment with an Efficient Integer Programming," IEEE Trans. Computer-Aided Design, vol. 20, no. 5, May 2001.

78. Ryan Kastner and Majid Sarrafzadeh, "An Exact Algorithm for Coupling-Free Routing," International Symposium on Physical Design (ISPD), April 2001.

79. J. Cong and D. Z. Pan "Interconnect Delay Estimation Models for synthesis and design planning". In proc. Asia and South Pacific Design Automation Conference, January 1999.

80. P. Chandrakasan et al., "Design of Portable Systems," in IEEE Custom Integrated Circuit Conference, 1994, pp. 259-266.

81. R. Conn, R. A. Haring, C. Visweswariah, "Noise Consideration in Circuit Optimization," in ICCAD98, 1998.

82. K. Soumyanath, S. Borkar, C. Zhou, and B. A. Boechel, "Accurate On-chip Interconnect Evaluation: A time-domain Technique," IEEE J. Solid-State Circuits, vol. 34, no. 5, pp. 623-631, May 1999.

83. HSPICE 2000.4 release document.

84. TEM waves at http://www.tpup.com/neets/book10

85. Lauren Hui Chen, Malgorzata Marek-Sadowska, "Closed-Form Crosstalk Noise Metrics for Physical Design Applications," in proc. The 2002 Design, Automation and Test in Europe Conference and Exhibition (DATE'02), pp. 1-8, March 2002.

86. Mohamed A. Elgamel and Magdy A. Bayoumi, "Interconnect Noise Analysis and Optimization in Deep Submicron Technology," IEEE Circuits and Systems Magazine, vol. 3, no. 4, pp. 6-17, 2003.

87. L. He, N. Chang, S. Lin, and O. S. Nakagawa, "An Efficient Inductance Modeling for On-Chip Interconnects," Proc. IEEE Custom Integrated Circuits Conference, May 1999, pp. 457-460.

88. P. Chen, Y. Kukimoto, C.-C. Teng, and K. Keutzer, "On Convergence of Switching Windows Computation in Presense of Crosstalk Noise," Proc. ISPD, 2002.

89. G. Zhong, H. Wang, C. k. Koh, and K. Roy, "A twisted Bundle Layout Structure for Minimzing Inductive Coupling Noise," Proc. ICCAD, 2000.

90. Y. Massoud, J. Kawa, D. MacMillen, and J. White, "Modeling and Analysis of Differential Signaling for Minimizing Inductive Cross-talk," Proc. DAC, 2002.

91. Xiaoning Qi, G. Wang, Z. Yu, and R. W. Dutton, "On-Chip Inductance Modeling and RLC Extraction of VLSI Interconnects for Circuit Simulation," Proc. IEEE Custom Integrated Circuits Conference, May 2000, pp. 487-490.

92. Switching   noise: Characterization   and   Analysis   Techniques, http://www.gigatest.com/pdfs/GTL81.pdf.

93. Mohamed A. Elgamel, Ashok Kumar, and Magdy A. Bayoumi, "Efficient Shield Insertion for Inductive Noise Reduction in Nanometer Technologies," IEEE Trans. on VLSI Systems, Vol. 13, No. 3, pp. 401-405, March 2005.

94. Mohamed A. Elgamel and Magdy A. Bayoumi, "Minimum-Area Shield Insertion for Explicit Inductive Noise Reduction," IEEE International SOC Conference, September 17-20, 2003, Portland, OR, USA.

95. Jiang-An He, Hideaki Kobayashi, "Simultaneous Wire Sizing and Wire Spacing in Post-Layout Performance Optimization," ASP-DAC, pp. 373-378, 1998.

96. Tianxiong Xue, Ernest S. Kuh, and Dongsheng Wang, "Post Global Routing Crosstalk Synthesis," IEEE Trans. Computer-Aided Design, vol. 16, no. 12, December 1997.

97. R. Arunachalam, E. Acar, and S. R. Nassif, "Optimal Shielding/Spacing Metrics for Low Power Design," IEEE annual symposium on VLSI, ISVLSI 2003, pp. 167-172, 2003.

98. Prashant Saxena and C. L. Liu, "A post processing Algorithm for Crosstalk-Driven Wire Perturbation," IEEE Transactions on Computer-Aided Design of ICs and Systems, vol. 19 pp.691-702, June 2000.

99. Mohamed A. Elgamel, Sumeer Goel, and Magdy A. Bayoumi, "Noise Tolerant Low Voltage XOR-XNOR for Fast Arithmetic," 2003 Great Lakes Symposium on VLSI (GLSVLSI2003), April 28-29, 2003, Washington D.C.

100. J. Cong, D. Z. Pan, and P. V. Srinivas, "Improved Crosstalk Modeling for Noise Constrained Interconnect Optimization," Proc. the ASP-DAC 2001, Asia and South Pacific Design Automation Conference, pp. 373 –378, 30 Jan.-2 Feb. 2001.

101. Mohamed A. Elgamel and Magdy A. Bayoumi, "An Efficient Minimum Area Spacing Algorithm for Noise Reduction," 10th IEEE International Conference on Electronics, Circuits and Systems, December 14-17, 2003, Sharjah, United Arab Emirates.

102. T. Gao and C. L. Liu, "Minimum Crosstalk Channel Routing," IEEE Trans. Computer-Aided Design, vol. 15, pp. 465-474, May 1996.

103. T. Yoshimura and E. S. Kuh, "Efficient Algorithms for Channel Routing," IEEE Trans. Computer-Aided Design, vol. CAD-1, pp. 25-32, Jan. 1982.

104. D. Sylvester and K. Keutzer, "Getting to the Bottom of Deep Submicron," ICCAD, pp. 203-211, 1998.

105. Rafael Reif, Andy Fan, Kuan-Neng Chen, and Shamik Das, "Fabrication technologies for three-dimensional integrated circuits," In Proc. IEEE Intl. Symp. On Quality Electronic Design, pages 33–37, 2002.

106. Said F. Al-sarawi and Derek Abbott. 3D VLSI packaging technology, http://www.eleceng.adelaide.edu.au/Personal/alsarawi/packaging_www.html.

107. Sridhar Krishnan, Young-Gon Kim, KM Bang, "A 3-D Stacked Package Solution for DDR-SDRAM Applications," Twentieth Annual IEEE Semiconductor Thermal Measurement and Management Symposium, 2004, 9-11 Mar 2004, pp.64–69.

108. J. Demmin, D. Baker, and W. Zohni, "Stacked chip scale packages: manufacturing issues, reliability results, and cost analysis," IEEE Symp. Electronics Manufacturing Technology, pages 241–247, 2003.

109. Matrix Memory, http://www.matrixsemi.com/

110. Semiconductor International, http://www.reed-electronics.com/semiconductor/article/CA621799.html

111. Keith D. Gann, Neo-Stacking Technology, http://www.irvine-sensors.com/pdf/Neo-Stacking%20Technology%20HDI-3.pdf

112. Terrazon Semiconductor, http://www.tezzaron.com

113. Cast, http://www.cast-inc.com

114. Embedded.com, http://www.embedded.com//showArticle.jhtml?articleID=163700022, 05/22/05.

115. K. Bazargan, R. Kasmer, and M. Sarrafzadeh "3D Floorplanning: Simulated Annealing and Greedy Placement Methods for Reconfigurable Computing Systems." Journal of Design Automarionfir Embedded Systems, 2000

116. H. Yamazaki, K. Sakanushi, S. Nakatake, Y. Kajitani, "The 3D-packing by Meta Data Structure and Packing heuristics," IEICE Tram Fundamentals, pp 639-645 April 2000.

117. P. H. Shiu, R. Ravichandran, S. Easwar, and S. K. Lim, "Multi-layer Floorplanning for Reliable System-on-Package," The 2004 International Symposium on Circuits and Systems, May, 2004, pp. V-69 - V-72.

118. Jason Cong, Jie Wei, and Yan Zhang, "A Thermal-Driven Floorplanning Algorithm," IEEE/ACM International Conference on Computer Aided Design, 2004, ICCAD-2004, 7-11 Nov. 2004, pp. 306–313.

119. J. M. Lin and Y.-W. Chang, "TCG: A transitive closure graph based representation for non-slicing floorplans," Proc. of ACM/IEEE Design Automation Conference (DAC-2001), pp. 164-769, Las Vegas, NV, June 2001.

120. Stefan Thomas Obenaus and Ted H. Szymanski, "Gravity: Fast placement for 3-D VLSI," ACM Trans. Design Automation of Electronics Systems, 8(3):298–315, July 2003.

121. Balakrishnan, Karthik Nanda, Vidit Easwar, Siddharth Lim, and Sung Kyu, "Wire Congestion And Thermal Aware Global Placement For 3D VLSI Circuits," Georgia Institute of Technology, CERCS;GIT-CERCS-04-17, http://hdl.handle.net/1853/102

122. Jacob R. Minz, Mohit Pathak, and Sung Kyu Lim, "Net and Pin Distribution for 3D Package Global Routing," date, vol. 02, no. 2, p. 21410, Design, 2004.

123. Jacob R. Minz and Sung Kyu Lim, "Layer Assignment for Reliable System-on-Package," Asia South Pacific Design Automation Conference, Jan 2004.

124. Tsung-Yi Ho, Yao-Wen Chang, Sao-Jie Chen, and Der-Tsai Lee, "Crosstalk- and Performance-Driven Multilevel Full-Chip Routing," IEEE Transactions on Computer-Aided Design of Integrated Circuits and Systems, vol. 24, no. 6, June 2005, pp. 869–878.

125. D. Y. Seo and D. T. Lee, "On the complexity of bicriteria spanning tree problems for a set of points in the plane," PhD dissertation in Northwestern University, Evanston, IL, 1999.

126. Jacob Minz, Sung Kyu Lim, and Cheng-Kok Koh, "3D Module Placement for Congestion and Power Noise Reduction," ACM Great Lake Symposium on VLSI, pp. 458-461, 2005.

127. M. Rewienski and J. White, "A trajectory piecewise-linear approach to model order reduction and fast simulation of nonlinear circuits and micromachined devices," IEEE Trans. on Computer-Aided Design of Integrated Circuits and Systems, pp. 155–170, 2003.

128. Brent Goplen, Sachin Sapatnekar, "Thermal Via Placement in 3D ICs," ISPD'05, April 3-6, 2005.

Index

About the Authors

Mohamed A. Elgamel received the B.Sc. degree in computer science from Alexandria University, Alexandria, Egypt, in 1991, the M.Sc. degree in computer engineering from Arab Academy for Science and Technology, Alexandria, Egypt, in 1998, the M.Sc. and the PhD degrees in computer engineering from University of Louisiana at Lafayette, Louisiana, USA in 2000 and 2003.

Currently, he is a research faculty at The Center for Advanced Computer Studies, University of Louisiana at Lafayette, USA. His research interests include CAD tools, high-level synthesis, noise aware layout, VLSI circuit design, and architectures for digital video processing.

He received the 2002 Richard E. Merwin Student Scholarship for demonstrating outstanding involvement in an IEEE-CS and excellence in academic achievement. He was the first place winner in "the student paper contest" in the 46th IEEE International Midwest Symposium on Circuits and Systems, (MWSCAS'2003). He also received the third place winner twice at University of Louisiana, IEEE Computer Society, Student Paper Contest.

He was the president of the IEEE Computer Society student chapter at the University of Louisiana at Lafayette and he served on the program committees of several conferences. He is a senior IEEE member.

Magdy A. Bayoumi is Director of The Center for Advanced Computer Studies (CACS) and Department Head of the Computer Science Department at the University of Louisiana at Lafayette (UL Lafayette). He is also the Edmiston Professor of Computer Engineering and Lamson Professor of Computer Science. Dr. Bayoumi has been a faculty member in CACS since 1985. He received the B.Sc. and M.Sc. degrees in Electrical Engineering from Cairo University, Egypt; M.Sc. degree in Computer Engineering from Washington University, St. Louis; and the Ph.D. degree in Electrical Engineering from the University of Windsor, Canada.

Dr. Bayoumi's research interests include VLSI Design Methods and Architectures, Low Power Circuits and Systems, Digital Signal Processing Architectures, Parallel Algorithm Design, Computer Arithmetic, Image and Video Signal Processing, Neural Networks and Wideband Network Architectures. Dr. Bayoumi is leading a research group of 15 Ph.D. and 10 M.Sc. students in these research areas.

He has graduated 25 Ph.D. and about 100 M.Sc. students. He has published over 200 papers in related journals and conferences. He edited co-edited and co-authored 5 books in his research interest. He was the guest editor of three special issues in VLSI Signal Processing and co-guest editor of a special issue on "Learning on Silicon." He is co-guest editor for two special issues on Video Architectures and Architectures for Machine Perception.

Dr. Bayoumi is the recipient of the 2003 IEEE Cts. and Systems Education Award.

He was the vice president for technical activities of the IEEE Circuits and Systems (CAS) Society, where he has served in many editorial, administrative, and leadership capacities. He was elected to the BoG (1996). He is one of the founding members of the VLSI Systems and Applications (VSA) Technical Committee (TC) and was the past chair. He is the chair (and founding member) of the Cts. and Systems for Communication TC. He was one of the founding members of the Neural Network TC. He is a member of the Multimedia TC. He has been on the technical program and organizing committees for ISCAS for several years. He has also organized several special sessions and workshops at this conference. He was a co-organizer and co-chair of a forum on MEMS in ISCAS'95.

He is a member of the steering committee of the IEEE Midwest Symposium on Circuits and Systems (MWSCAS). He was the general chair of MWSCAS'94 and 2003.

He was an associate editor of the IEEE Circuits and Devices Magazine, IEEE Transaction on VLSI Systems, IEEE Transaction on Neural Networks, and IEEE Transaction on Circuits and Systems II.

He was the general chair of the 1998 IEEE Great Lakes Symposium on VLSI, the IEEE VLSI Signal Processing Workshop 2000, and the 2004 IEEE International Symposium on VLSI. He represented the CAS Society on the IEEE National Committee on Engineering R&D policy, 1994, the IEEE National Committee on Communication and Information Policy, 1994, and the IEEE National Committee on Energy Policy, 1997 and 2004.

Dr. Bayoumi serves on the IEEE ASSP Technical Committee on VLSI Signal Processing. He has been a member of the Technical Program of the IEEE VLSI Signal Processing Workshop, the International Conference on Application Specific Array Processors, and the Computer Arithmetic Symposium. He was the general chair of the Workshop on Computer Architecture for Machine Perception, 1993 and he is a member of the Steering Committee of this workshop.

Dr. Bayoumi served on the Distinguished Visitors Program for the IEEE Computer Society, 1991-1994 and the Cts and Systems, 1999-2001. He is the faculty advisor for the IEEE Computer student chapter at UL Lafayette. He won the UL Lafayette 1988 Researcher of the Year award and the 1993 Distinguished Professor award at UL Lafayette. He is an IEEE Fellow.

Dr. Bayoumi served on the technology panel and advisory board of the US Department of Education project, "Special Education Beyond Year 2010," 1990-1993. He was the vice-president of Acadiana Technology Council. He was on the organizing committee for Acadiana's 3rd Internet Workshop, 1999. He gave the keynote speech in "Acadiana Y2K Workshop," 1999.